Best,
Ed Millis

TI,
the Transistor,
and Me

or

My Dis-Integrated Circuit Through
Texas Instruments

by edwin graham millis

© 2000
Ed Millis Books

Edwin Graham Millis
9405 Forestridge Drive
Dallas. Texas 75238

ISBN 0-9709463-2-5

Cover. with apologies to Jack Kilby. by my darling daughter
Bev Haskin. The art degree pays off.

Fourth Edition 8 May 2001

Manufactured in the United States of America
by Excel Digital Press. Inc.. Carrollton. Texas

To all the TIers...

both here and gone, who made my career such a joy. And also miserable, exciting, frustrating, fulfilling, and just plain fun!

Preface

When I retired from TI on the last day of March 1989, I didn't realize the quantity of memories about my days of employment that were hidden in my brain. But I liked the stories and wanted to preserve them, so I soon began carrying a tape recorder in my car to "jot down" key words to remind me of the incidents. And within a year or two I found myself recalling stories that I had already recalled, so I quit, mostly, and made a swell organized list of things that happened to me, or that I did, or that I might have done if I'd been smarter. The final list was about 200 items. Better judgment kept me from using about a third of them, generally on the basis of being just too dumb. Things that might sound funny in a crowd after a couple of drinks may not stand the translation to a piece of paper. At least not the way I write.

And that was the basis of this book. That, and weeks of digging through the documents I had saved from my years at TI. And then, the most fun of all, calling old friends, some of whom I hadn't talked to in years, to straighten out some of my so-called facts. What a great side benefit!

I soon found that I could not write this book without using the language of my profession and for that I apologize to those who do not understand it. You should have seen the first draft before I took most of it out. So for those of you who might think a "p-n junction" is the intersection of two urinals, forgive me. I truly don't think it will affect the narrative of my thin trail through Texas Instruments. But for those fortunate enough to have worked in that embryonic era, I hope the jargon brings back good memories of the time.

Texas Instruments was a unique place to work in the early days of transistors and integrated circuits. As engineers we were given a freedom seldom found in industry. The future of semiconductors was only dimly visible to us, and there were few guideposts on the way. Sometimes we went in the right direction and sometimes not. The good news was that management couldn't tell the proper direction either. But whatever we did, we did it at a dead run, and if it was wrong and we fell down, we got up, shook ourselves. turned around and ran off again on a better path. Was it the golden age of engineering? Possibly not, but it was an immensely satisfying time for my creative urges.

I suppose I could complain about the many unpaid overtime hours I worked, and that sometimes, I'm sorry to say, the family came second. But the excitement of those times was overpowering, and looking back I can only wonder how we as people survived. But survive we did, and I hope that in some small way we put a couple of stones in the foundation of a great company, Texas Instruments.

And for those of you picky ex-TIers who find fault with a story or two, and just *know* that it happened differently than my version, tough shit. My mind is perfectly clear after all the years of filtering the facts, and I'll stick by what I wrote. As for you? Write your own TI book. I'd love to read it.

Acknowledgments

My heartfelt thanks to my dear friend Shirley Sloat who spent many hours with her red pen in an attempt to keep me from embarrassing myself in print. It's not her fault if I do. Lord knows, she tried.

And thanks to my children, Bev, Susan and David, and their families for encouraging me to write this book. I hope they find a few new stories in it. And some extra thanks to Bev for the terrific cover art and the editing help. Actually, thanks to everyone who encouraged me in this endeavor, which was nearly everybody I know. Now see what you've done!

Contents

Chapter One

Bob Kelly greeted me on the morning of Tuesday, September 2, 1950, and welcomed me to Geophysical Service Incorporated. He led me down the main hall of the Lemmon Avenue plant, GSI's newly completed and only Dallas facility, up the stairs and into the library. Maybe, I thought, I was going to get to read up on the history of the company or something. Bob didn't miss a step as he passed through the library, raised the window, and crawled out on the roof. He turned around and said, "Come on out. Here's where you're going to work."

I knew that I was going to be checking out Navy airborne radar systems but somehow the thought that it would be on the roof didn't occur to me. But I followed Bob out of the window and over to a tent. Exactly how they pitched a tent on the roof without causing serious leakage below I didn't find out, but in the tent were a couple of technicians and a couple of APS-31 radars.

I, Ed Millis, recent graduate of The Rice Institute, figuratively clutching a Bachelor's Degree in Electrical Engineering to my scrawny breast, had just been officially employed on August 31, 1950, as Employee Number 9345 at GSI, soon to become Texas Instruments. I could not have picked a more fortuitous bit of employment in the whole world. Much as I'd like to take credit for incredible foresight, it was not cleverness on my part that brought me here, but the idea of making two bits an hour more than I was making wiring seismograph trucks for National Geophysical down the street. That, and being on a bowling team. It was a hell of a deal, rooftop or not. And just to add to the pluses, I didn't have to come to work until Tuesday, the 2nd of September, because of Labor Day. My first day of employment was a holiday.

I had taken the only job I could find in Dallas after I graduated in June of '50. I was working as an electrical technician for National Geophysical on Lemmon Avenue across from the old Love Field passenger terminal. It was a big letdown after the life-or-death struggle I had just been through at Rice. Technical jobs were scarce to none with only three in my class of EE's having jobs at the end of school, and two of those were General Electric trainee openings. So I started at National for a buck an hour wiring up

1

seismic trucks. I could do that about as well as anyone, and so had begun to settle into a not-very-exciting routine of working eight hours a day and making eight dollars. Less taxes, that is.

It sounded like a good idea when long-time friend Joe O'Connor suggested I put an employment application in to Geophysical Service. It sure couldn't be any less exciting than where I was, and the equipment he was working on sounded a lot more interesting.

Joe was an electrical technician at GSI and a good one. He stuttered badly but made up for it by having all the latest and funniest jokes to tell, no matter how long it took. He said they were building military equipment like radars and submarine detection systems and it was a good place to work. I stopped by and filled out an application and waited.

Joe bugged me regularly. "Have you h-h-heard from GSI y-y-yet?"

"Joe," I replied," I called there again today and they still can't find my application. Got any good ideas?"

"Yeah," said Joe. "Fill out another one." So I did.

This conversation was repeated the following week. I filled out a third application for employment. I was beginning to think they didn't like me. But this time it worked. I called Joe. "Joe, they told me to come back in Friday and they'd get me a physical and I could start immediately. Hot damn!"

"Th-th-th-that's great!" said Joe. "And it's a g-g-good thing t-too, since I already s-s-signed you u-up on my bowling t-t-team. You b-bowl next Th-Th-Thursday night." Sounded fine to me.

Just about then, Mary Ruth Street, head of the GSI Engineering Department secretarial pool and personal secretary to Jim Wissemann, Chief Engineer, was standing at the office bulletin board. She was looking over the bowling league sign-up sheets to see who her team members would be for the new season. She had vocally complained to Joe O'Connor, team captain, that she was tired of being the only woman. But Glory Be! He did listen to her! At last! Edwina Millis was going to be on their team!

And so once a week that fall the TI Bowling League became my evening entertainment. Mary Ruth was disappointed that Edwina Millis turned out to be just plain Ed, just another man, and an engineer at that. But we hit it off famously, kind of. Our first date, which was taking her home after bowling, began by my borrowing a quarter from her to buy gas for the car. As I recall it went downhill from there. But, to make a long and pleasant story shorter, we were married on the 18th of October 1952. Bowling can be a life-changing experience.

It was a minor inconvenience to work on APS-31 radars in a tent on the roof, and it didn't bother me in the least. I was already a believer in the "it all pays the same" theory of employment. If they wanted me to stand on my head and check out radars, fine. Just clear me a little more space so I can get up. And besides, the rooftop tent business didn't last long. We soon moved into the spanking-new Radar Test Tower.

Putting the radar boxes together and testing them was a lot easier and more comfortable in the checkout tower in the rear of the Lemmon Avenue building. It was a large square room perched on top of the second floor above the manufacturing area with an unobstructed view of the city from the northeast to the southeast. The room was serviced by an elevator to reduce hernias, and one of the large floor-to-ceiling windows was acrylic plastic to shoot the radars through.

For a memorable period in the first months of 1951 I checked out radars on the graveyard shift. For the lucky ones of you who don't know the term "graveyard shift," it was an aptly-named work period beginning at midnight and lasting until 8:00 in the morning. The good news was that it paid a few cents more per hour for the insult, and the bad news was that it was from midnight until 8:00 in the morning.

Working the graveyard shift while dating Mary Ruth was a handicap. It required timing the date to end about 11:45 in the evening and then promptly hauling my body to the plant to check out radars. Not exactly the ideal way to end an evening of courting, but Mary Ruth understood and put up with it. Besides, I was making a dollar and a quarter plus the fifteen-cent shift premium. That's over eleven bucks a day! Less taxes, that is.

The good thing about ending work at 8:00 a.m. was having the mornings free. The key to survival was getting enough sleep before the next midnight rolled around. I can remember times during the winter of '51, about three in the morning, being sleepy beyond belief and catching a few quick winks while curled up around a nice warm radar modulator unit. One of the little-appreciated qualities of vacuum tube equipment—they were warm in the winter.

But I put my free mornings and weekends to good use in the winter of 1951. I had taken flying lessons in the summer of 1950, and had been flying every chance I could squeeze in, finances permitting.

I was still living with my parents at the time I decided to learn how to fly, and they were less than thrilled with the idea that I'd be sailing around up in the air without a parachute. Ploy Number One worked just fine. Oh. well, I said, I guess I'll buy a motorcycle instead...

On August 27, just before I began work at GSI, I started flying lessons at the Highland Park Airport, near the intersection of Valley View and the future Central Expressway. Julius Hudson, owner and operator of the airport, was my instructor. I don't think any conclusions should be drawn from this. but I apparently was his last student as he gave up teaching immediately afterwards.

My cherished flight log shows I must have liked the first flying lesson on the 27th of August. because I took another lesson on the morning of the 28th. And then I took another in the afternoon, followed by a fourth after my first day of work at GSI, the 2nd of September. I was getting the hang of it. I must have had some money saved up since dual instruction in the Piper J-3 *Cub* cost an incredible $9.00 per hour. It was easy to blow a large piece of a day's pay at the Highland Park Airport. The lessons were usually 30 minutes. clocked by a stem-winder dollar watch that was fastened to the instrument panel of the *Cub*. After I propped the engine, Mister Hudson would set the watch to 12:00. wind it up a turn or two, and I'd climb in. We were ready to go flying.

And on the second of October. I achieved the other thrill that only occurs once. We had been shooting practice landings when Mister Hudson shouted over his shoulder to me. "Taxi over near the gas pumps and let me out. Then take it around by yourself." Mr. Hudson always shouted in the airplane to be sure you heard what he had to say. I heard. although in a state of disbelief, but I taxied over and stopped. engine idling. and he got out.

Anyone who has soloed will remember the excitement of that first take off. circuit of the airport. and landing. After I'd taken off and made the two left turns at just the right altitudes and entered the down-wind leg of my airport flight pattern, I remember thinking, "What the hell am I doing up here by myself?" The front seat of the Cub looked so... so *empty*. But Mister Hudson was right in believing I wouldn't wreck his training plane, and so after three hours and ten minutes of dual instruction, I soloed.

Performing the final checkout and Navy acceptance tests on the APS-31 radars had become routine to me by the spring of 1951 and the night shift was getting to be a real drag. I was delighted to find that I was to be transferred to the ASQ-8 "MAD" checkout on the day shift. In military equipment talk, MAD meant Magnetic Anomaly Detector. This equipment was carried by both Navy aircraft and Navy blimps and was used to locate submerged enemy submarines.

The fleet submarines of that time were essentially big pieces of iron and steel, and as you might expect, screwed up the earth's magnetic field in their vicinity. They screwed it up enough that the ASQ-8 could detect this magnetic warpage, a.k.a. "anomaly," many hundreds of feet away. This was a useful method of finding a submarine hiding under the surface of the ocean.

The ASQ-8 was an immensely interesting piece of equipment. I was ready to go find out about it and get back to the daylight hours.

But first, one little chore. I had to become a Deputy Sheriff. I had to become a Deputy so I could carry a gun. I had to carry a gun because the ASQ-8 was classified "Military Confidential," and any transport outside of our plant had to be done with an armed guard. I was to become an Armed Senior Electrical Technician who could guard the equipment to and from the remote test site near the town of Grapevine.

After my last graveyard shift with the APS-31s, I drove downtown to the Dallas Sheriff's headquarters to be deputized. I was shuffled around and finally taken in hand by a trustee of the Dallas County Jail to be fingerprinted. The trustee, noting my unshaven appearance and dirty khakis as he inked up my fingers, asked, "What are you being booked for?" How rude. But I didn't care. I was getting back on the day shift with the ASQ-8. Sayonara, APS-31!

Chapter Two

My new job with the ASQ-8 was basically the same as it had been with the APS-31—assemble electronic boxes into a complete system and see if it runs properly and meets the Navy's specifications. But the ASQ-8 MAD gear was a technically challenging piece of equipment. It could detect a change in the earth's magnetic field strength of one part in a hundred thousand. And, as you may have noticed, the earth's field is not particularly strong. A special calibrated magnet, like strong enough to pick up a few paper clips, was used to test the sensitivity of the ASQ-8. This magnet rotated thirty feet from the detector would cause a full-scale signal on the chart recorder. On a good day an ASQ-8 could find a 1950's vintage fleet submarine 2000 feet away.

To get this incredible sensitivity required quite a collection of electronic and mechanical doohickies. Today the whole of the electronics could be put on a few chips, but in the vacuum tube days it took four chasses, each weighing twenty or thirty pounds, plus a control box and the detector assembly. It was a substantial system.

Checking out this system to see if it met the Navy's requirements required some special doings also. Sensitive to magnetic surroundings, like power lines and moving automobiles, the test location was out in the boondocks, south of Grapevine Texas. The site is now part of the DFW International Airport. The Field Camp, as it was called, was constructed in the center of a large acreage of rolling pastureland, far from any roads or power lines. The only facility was an antique party-line crank telephone mounted on a pole. It may have been the last one in Texas, but it was our lifeline to the Lemmon Avenue plant. The Field Camp had the extremes of technology.

The ASQ-8 Field Camp was self-sufficient. It had several electrical generators in a prefab garage to generate both the standard house-current to run the lights and such, and also the special voltages and frequencies needed by the ASQ-8. Adjoining the generator garage was the parking place for the twenty-eight foot house trailer that was hauled back and forth from the Lemmon Avenue plant daily. This house trailer was equipped like a checkout lab, with workbenches and test set-ups. ASQ-8s ready for checkout were loaded into the trailer, taken to the Field Camp, and tested.

After sign-off by an on-site Navy Inspector, they were trucked back to Lemmon and packed for shipment. Later a one-room frame house equipped for testing was built. and just us chickens and the equipment shuttled back and forth.

About a hundred and fifty feet from the trailer were three test shacks where the magnetic detector heads were set up. All the serious magnetic testing was done in these little non-magnetic wooden houses. They were built with no magnetic materials—that is, no steel flashing. hinges, or nails. Everything was either wood or aluminum. The cradle that held the detector was similarly all wood with glued dowel joints and no metal fittings.

One test required a person go down to the test shack and "rock the cradle" to simulate the action of a moving airplane on the detector. This person had to strip himself of all magnetic objects, so he would frequently be seen holding up his pants with one hand while rocking the cradle with the other. A steel buckle on your belt or a paper clip in your pocket would mess up the signals big-time. Fortunately, zippers were brass.

I worked the day shift. the night shift. and then the night-and-day shift when we were trying to meet a delivery schedule at the end of the month. After a big push to get systems out at the end of the month, we would all breathe a big sigh of relief for having made our schedule. take a couple of days off to recuperate. and then start the next month already behind. I can remember trying to catch a few winks on top of the hard workbenches on cold winter nights at the Field Camp.

The converted house trailer had a kerosene "central heat" that might have been sufficient in a sheltered cluster of other buildings. but on a barren hill in a Texas winter it was less than marginal. It got cold in the trailer. and it behooved us to keep an eye on the kerosene level in the tiny tank. If it ever ran out. you risked being frozen to death before you could get it refilled and lit. Several times a day one of us would go outside to the kerosene barrel. pump out a few gallons, and transfer it to the trailer heater tank.

I had gone out on a bitter cold day to perform this function and was standing near the trailer door with my can of kerosene when I heard *Bang!* from inside. At the same time a hole appeared in the aluminum skin of the trailer a foot or two to my left. I put the kerosene down and opened the door to see what was going on.

7

The Navy inspector was standing with a smoking pistol in his hand, eyes big as saucers, and unable to get anything out but, "Uh, uh, uh...."

He had taken the automatic that we carried with the equipment out of its holster, "unloaded it," and pulled the trigger. He shot through the tabletop and out the side of the trailer. Fortunately for me, it was not where I was standing.

After the inspector got his voice back and nerves calmed, we jointly decided that this was merely a minor incident that didn't need to be reported. Besides, I needed him to sign off a couple of ASQ-8s later in the day. Having his ass arrested and thrown in jail for being a stupid idiot would have been satisfying but unproductive. So he patched up the hole in the table top while I pounded the splayed aluminum shut on the outside of the trailer. Good as new and totally undetectable. About thirty years later I ran into Bob Kelly, head of the Field Camp operations at the time, and confessed. And then he confessed he had figured it all out after seeing the repairs and had also decided not to make an issue of it. Particularly since no one had been killed.

The night shifts were worse than those back on Lemmon Avenue. The Field Camp was a lonesome place even though we were required to work in pairs at least. But the nights were long and frequently boring, although not always.

The "not always" was the night Clarence Elmore and I were checking out systems and noticed the room lights flickering. We ambled out to the generator garage and stuck our heads in. A haze of smoke and a small fire on an inside wall greeted us. The control system for the biggest generator was aflame. Some clever person had nailed the control components to the wooden siding and then wired them to the generator. This didn't exactly meet the electrical codes of Texas. Or Nicaragua for that matter.

One of us hit the kill switch on the generator and the other tripped the main disconnect breaker. All the lights in the Field Camp went out as Clarence and I scrambled to minimize the damage. Clarence yelled, "I'll get the fire extinguisher!" and headed out the door in the dark, only to trip and fall. Cursing, he got up and I heard his running footsteps fade away into the night. Not exactly knowing what to do, I felt my way over to the fire on the wall and began trying to blow it out. It was the only thing I could think of. It's amazing how hard you can blow if you're trying to keep the Field Camp from burning down on your shift. And I blew the fire out.

Clarence soon reappeared at the door with an extinguisher and a flashlight, but I was able to gasp out that everything had been taken care of, thank you. Further inspection showed that the object that he had fallen over on the way out was the other fire extinguisher.

Clarence and I were good friends and once got into a little sartorial competition. We decided to have an Ugly Tie Contest. Just the two of us. Judging was to be done by our fellow engineers, none of whom had the least sense of taste in clothing, much less ties.

Since I could sew to some slight degree, I headed to a cloth store to see what I could find for raw material. My theory was that tie makers wouldn't purposely make ugly ties. Accidentally, yes, but not on purpose. I, however, was going to make an ugly tie on purpose, which should be an order of magnitude uglier than the accidental kind.

The clerk at the cloth store was nonplused by my request "to see some ugly cloth." However, after a brief discussion of my quest, he got into the swing of things and soon found a candidate for me. In his words, "Now here's some *really* ugly cloth." And boy, was it. Purple and black with a touch of lime green. I bought a half a yard and thanked him for his assistance.

The contest was an anticlimax. I sewed up a tie from this really ugly cloth and won hands down. Clarence thought his hand-painted hula girl tie was ugly? Clarence don't know ugly.

Finally it was my turn to go into the "field," the Naval Air Stations, and work on ASQ-8s. It was with a mixture of high anxiety and eager anticipation that I gathered my tools and prepared to visit the Navy blimp headquarters at the Lakehurst Naval Air Station in New Jersey. This trip would be a first for me in several ways. Even though I now had dozens of hours of light plane stick time to my credit, I had never flown on a commercial airliner. And I had never been east of Terrell, Texas, which is only fifteen miles away. I had been to California, but never east, and especially never to New York City, the first stop on my trip.

The flight to New York City in an American Airlines DC-4 took four or five hours as I recall, and was a lot like being put into a

9

tin can and vibrated and deafened. It was interesting because I was interested in airplanes. It was even interesting when we circled New York that night in a storm for what seemed like an hour. I understood what vertigo was when an illuminated "Consolidated Edison" sign went by my airplane window at an impossible angle and speed. I just thought I knew which way was up. But we landed safely and I found my way to a hotel in downtown New York, near the Port Authority Bus Terminal. You understand, of course, that every move I made was carefully choreographed on a piece of paper clutched in my clammy hand. This was not a time for creativity.

Early the next morning, I walked around the corner to the Port Authority Bus Terminal and took a seat on the bus to Lakewood, New Jersey. At Lakewood, I took the only taxi in town to the Lakehurst Naval Air Station. It was a familiar drive for him since most of his customers were to and from Lakehurst.

Lakehurst was always a lot of fun to visit. The Blimp Navy, or "the Poopie-Bag Navy" as the sailors called it, was a world of its own. I made a lot of trips there and was never bored with the blimps or the people that worked with them.

Several of the big hangers at the Lakehurst Naval Air Station were built in the 1930s. One of these was Hanger Number One, the biggest on the base, and it was huge. It could hold nine of the "fleet" blimps—the "K Class" airships I later flew in—with three rows of three end-to-end blimps. The K ships were the standard Navy blimps that were used for ocean patrol duty. They seemed incredibly large and they each held 425,000 cubic feet of helium to make them light enough to fly. They were powered by two nine-cylinder radial engines mounted just outside the "car" which is what you rode in and flew it from. The car was built just slightly better than flimsy, since weight was crucial on LTA aircraft. LTA means "Lighter Than Air" in case you weren't in the Navy at that time. HTA, similarly, means Heavier Than Air, like real airplanes and such.

There were a number of seats in the K ship's car, kind of like those in commercial airliners but much lighter. Observers could sit near the rear and look out huge plastic windows that completely surrounded the car. The pilot, or captain of this ship of the air, sat on the left side at the front. His controls were the engine's throttles and lots of funny cables he could pull and lock to control the pressure in the helium cell and the air-filled ballonets (French pronunciation).

10

He also had a wheel by the right side of his seat, much like a wheel on a wheelchair. This swell wood-rimmed wheel controlled the elevators which made the blimp go up and down. What he didn't have was the wheel for the rudder, which made the blimp turn right and left. He wasn't quite the complete captain of this strange ship, because the copilot got to do the rudder.

The copilot sat in the right front seat with an even bigger wood-rimmed wheel directly in front of him. This wheel resembled a wheel from a sailing ship, except that it was thinner and didn't have the short spokes on the outside. Just a pretty polished wooden wheel. It would seem smart for the pilot to stay on good terms with the copilot since his cooperation was vital if he had actually chosen some specific place to go.

Later that first day at Lakehurst I got a ride in a K ship. They wanted to take me, the ASQ-8 "expert," out beyond the city of Lakewood and out over the ocean and run the equipment. They knew of several wrecked ships that made good magnetic signals for testing. I climbed the thin ladder into the car of the airship and took a seat. I don't recall anyone fastening seat belts. I don't remember that there were any. But nothing happens very fast on a blimp so they really weren't needed.

The engines were started and the blimp, nose fastened to a tractor-driven mast, was towed out to an open space away from the hangers. It was disconnected from the portable mast and a cluster of sailors held "lines," i.e. ropes, to keep the unwieldy airship in position. At the signal of "Up Ship!" (and you can guess how far back *that* command dates) the sailors dropped the lines and the engines were throttled up to full power. The noise inside the car was deafening. It looked like the propellers, one on either side, cleared the car windows by about an inch, and the whole car vibrated and rattled like a Model T Ford at full throttle on a rough road. The blimp ran along the ground on its single doughnut tire for a few dozen yards and then the nose pointed skyward steeply enough to make an airplane pilot blanch. This was, in my brief light plane flight training, a guaranteed way to stall, spin in, and kill yourself. A blimp pilot later told me that their favorite trick was to take a Navy fighter pilot for his first blimp ride, and after hauling the nose up at an incredible angle on takeoff, chop the throttles about fifty feet off the ground. The fighter pilot knows they're all going to die in a fiery

11

crash, but the blimp just kind of rolls back level and sits there. After reviving the Navy pilot, they throttle back up and continue their takeoff.

Strangely enough, the LTA, "Lighter Than Air" it says here, blimps aren't. They're actually heavier than air. The K ships normally weighed in about 4000 pounds heavier than the helium could lift standing still. But as soon as that huge curved gas bag got moving a little and tipped up for some angle of attack, it generated plenty of lift to make the blimp fly. So they can't actually stop in midair for very long, but they drift downward so slowly it seems like it.

My first blimp flight had another civilian employee on board, Jack Lawson. He worked for the Bureau of Aeronautics and was an expert on Magnetic Anomaly Detection gear, such as the ASQ-8. I was told by my boss to be nice to him since TI was trying to hire him to come to work in Dallas. Not to worry, I wasn't going to be ugly to anybody while airborne in a blimp.

We headed out from the New Jersey coast and made a bee- or maybe a blimp-line for the Barnagat Bay Light Ship. This ship, rigged like a lighthouse, was anchored off the coast near, as you might have guessed, Barnagat Bay, and formed a nice permanent landmark out in the ocean.

After doing whatever we did that day so long ago, we headed back. The captain of our airship (and it rolled and wallowed in the sky just like a big sailing ship on the ocean) called to me and told me to go forward and crawl down into the nose gunner's seat. He gave me a short and mysterious message as I passed by him, "Watch the chickens!" Well, okay.

I squirmed down into the cramped seat of the forward gunner's station and was rewarded with the best aerial view of the world bar none. After the momentary disappointment of not actually having a machine gun to play with, I enjoyed the scene immensely. And then I saw the chickens.

A White Russian community populated the coastal area around Lakewood, most if not all of whom raised chickens. Each chicken farm had a square fenced area, plainly visible from our low altitude, with a square chicken house in the center. I could see a lot of them that looked the same, like they only had one set of chicken farm plans or something.

As the blimp approached, the chickens would suddenly come out of their little chicken house at full speed. At first I thought it was a cloud of smoke, but it was a cloud of chickens running faster than you could believe. And they kept running straight into the chain-link fences. The chickens must have thought the blimp was the biggest hawk in the world. They have a serious fear of blimps for whatever reason, and I suspect that it's frequently fatal when combined with chain-link fences.

But the blimp pilots thought scaring chickens was a lot of fun. Besides, on this flight we were late getting back to the base and if we didn't hurry the landing ground-crew would be sent off to mess and we'd have to wait around up in the air until they had eaten. So we buzzed the chickens at full throttle, which put our blimp at about 80 knots. Not bad speed for shoving over four hundred thousand cubic feet of helium in a big cloth sack through the air.

Watching chickens strain themselves through fences from the nose gunner's seat was fine, but Jack Lawson was doing something even better—the captain let Jack sit in the right seat and fly the rudder on the way back. My God, I was so jealous! Hell, I could do that if they'd just let me. But Jack had seniority and wouldn't turn loose of the beautiful polished wood-rimmed rudder wheel. Damn.

On a later flight one of the crew pointed out to me the different color of the vegetation on the ground in one part of the Lakehurst station. It was where the *Hindenburg* had burned, and the heat of the fire or the debris from the blaze had somehow changed the soil enough to make it visible from the air. I remember thinking at the time that the disaster had happened a long time before, but actually it was only about fifteen years earlier, in 1937. They also told me that the Hindenburg had been docked in the immense Lakehurst Hanger Number One on one of its trips to the U.S., and it was so long that it stuck out the end. That I would like to have seen. And because the Navy never threw anything away that they might need later, the huge movable mast for mooring the Hindenburg was still at the station. Completely out of scale compared to the tractor-drawn masts in use with the blimps, it sat forlornly in a corner of the Naval Air Station property, slowly rusting away on its rail tracks.

My second trip to Lakehurst was even more exciting than the first. I returned from lunch on a dreary and rainy day and made my

13

way up the stairs to the electronics shop on the second floor of the hangar. As I strolled into the shop, every sailor's head jerked my way and the conversation stopped abruptly. The Chief jumped up from his chair and said, "We thought you were dead! We just heard from the Lieutenant that you had jumped from inside the top of the hanger and committed suicide!" I assured them that I hadn't. I later found out that a civilian employee about my size had decided to kill himself. Since it was a nasty day outside, he chose climbing the catwalks up to the top of the inside of the hanger and jumping to the concrete floor below. Apparently very effective, and he didn't get wet. A Lieutenant of my acquaintance had been walking nearby and had seen the body and decided it was that guy from Texas working on the ASQ-8s. I was glad to report that it wasn't me flattened on the concrete.

But it could have been, almost. One thing I had to do occasionally was climb to the top of the tallest, ricketiest, portable crank-up ladder the world has ever seen. These Victorian-looking wooden ladders were like the extension ladders on fire trucks, and could be extended upward into the clouds if you just kept cranking long enough. The whole apparatus rolled around on 1890 wooden wagon wheels, and if OSHA had been alive in the '50s, it would have been immediately burned for firewood. But you needed one to work on the DT-37 detector head. The detector head was mounted way high on the front of the blimp's envelope, and, as Mohammed, I had to go to the mountain. I learned to choke down my fears and do the job, but I was never comfortable thirty-five feet in the air on a shaky ladder.

Peripheral entertainment occurred irregularly at the Lakehurst Naval Air Station and I was privileged to see two of the most popular activities—graduation ceremonies for parachute riggers and the beginning of the balloon trip that all aspiring blimp pilots were required to take.

Graduation from the Lakehurst parachute riggers school was simple. You packed a parachute and then jumped out of an airplane with it. Graduation was strictly pass-fail. I'm pleased to report that all the riggers passed the day I saw them step confidently from a C-47 over the base grounds. It certainly brought the base to a halt when this happened. Everyone was outside watching the jump. They ought to be ashamed of themselves.

14

The balloon launches, which occurred only a couple of times a year, were a similar pass-fail test for blimp-pilots-in-training. Three or four helium-filled balloons, looking like the one in "Around the World in Eighty Days," would be launched at the same time. Each carried two would-be pilots, a battery-powered weather radio and an altimeter. The only controls they had were a dozen or so sacks of sand for ballast and a rope leading to a helium dump valve. The object was to stay aloft and alive for several hours, and come back down without completely destroying the balloon and its wicker basket. Yes, if wicker was good enough for Montgolfier, it was good enough for the US Navy. Whiling away the evening hours at the bar in the Bachelor Officers Quarters where I stayed on the base, I soon learned to ask the pilots, who were also whiling away the evening hours at the bar, about their balloon flights. They loved to tell the hair-raising tales, and, as it was always the highlight of blimp pilot training, the tales were always hair-raising.

In the fall of 1952 I finally got some sea duty with the US Navy. TI had gotten a contract to design and build an ASQ-8 test bench that could be used on board an aircraft carrier. The good news was that Roger Webster, our most excellent ASQ-8 engineer, would go with me and we would spend ten days on the aircraft carrier CVB-43, the USS *Coral Sea*, off the East Coast. The bad news was that we were leaving two weeks after Mary Ruth and I were married. From honeymoon to a ship with 3,000 sailors. Large bummer.

But Roger and I had an interesting life aboard the Coral Sea, living very nicely in the Junior Officer's Quarters and eating a lot of good food. We did our magnetic testing with no particular difficulties, and by and large enjoyed the different world of a US Navy ship at sea. I particularly liked to watch the flight operations when possible and, with a little help from a sailor, discovered the Flag Bridge. It was high enough above the flight deck to be reasonably safe, and was equipped with a waist-high steel plate that I was prepared to duck behind. All manner of new jet aircraft were out on this cruise to do their aircraft carrier acceptance tests, so I got to watch a lot of catapult takeoffs and tail-hook landings.

Roger developed a problem brought about by the steep shipboard "ladders." The foot on his artificial leg began to come loose. We were in the electronic shop far below decks when Roger asked a sailor if he could borrow a 5/8 socket, extension and ratchet.

15

They obliged and Roger, to their amazement, proceeded to take his leg off and tighten his foot.

I was really glad to get back home after that trip. Mary Ruth, with a bit of coaching, remembered who I was and we picked up where we had left off on our new life together.

Another active field service location for ASQ-8s was at the Key West Naval Air Station. It was a great place to go and getting there was half—well, a third—of the fun. A flight to the Miami International Airport from Dallas got you close, and then it was a short walk over to one of the two National Airlines gates. There you boarded a Lockheed *Lodestar* for the final hop to Key West. The *Lodestar* was a small twin-engine tail-dragger that went through WWII as the *Hudson* bomber. I don't think that National's plane had actually gone through WWII since there were no signs of gun turrets or a bomb bay, although it looked old enough.

A passenger would enter the rear of the plane, which was an easy step off the ground, and walk up the center aisle and find a seat. No problems about seatmates since there were only a single row of seats on either side of the aisle. If you went to the forward end of the plane, the last couple of seats required stepping over the main wing spar which happened to go through the cabin about knee-high. It was padded and tastefully upholstered and didn't detract a great deal from the décor.

The flight to Key West didn't take long and soon ended on the coral airstrip of the Key West Airport. Two people met the plane: the National Airlines employee who ran the airport by himself and the one and only Key West taxi driver. The "terminal" was a wooden shack with a tiny counter and a few chairs. I was interested in the steel cables that ran over the top of building to huge concrete cylinders on either side of the shack. I asked, and the answer was that it kept the terminal from being blown away in a hurricane. Simple but effective I suppose.

The taxi driver opened the cargo door on the *Lodestar* and unloaded the baggage onto a cart while the National employee and the pilot busied themselves connecting up an external battery or something. At least it had electric starters. After a few minutes, most of us had crowded into the rusty taxicab and headed into town. I went as far as the Submarine Base on the eastern edge of Key West. My normal accommodations while working at the Naval Air Station were in the Submarine Base BOQ (Bachelor Officers Quarters) for the grand price of a buck a night.

Somehow, at that time in the life of Texas Instruments, the idea of my actually renting a car to get around had never occurred to anyone. And of course in my semi-lowly position of Technician or Junior Engineer, I didn't rate near high enough to think of it myself. So I considered it was no big deal to hitch rides to and from the Naval Air Station daily, since I could usually find someone to pick me up. Occasionally I was reduced to thumbing by the side of the road to get to the base. That worked, too.

The BOQ was just fine to stay in, especially at this undemanding stage of my life. The rooms were just slightly seedy and had neither heating nor air-conditioning. which is okay in Key West, but they were equipped with nice manually-operated windows. The Sub Base also had a place to get a bite to eat, and of course there was a bar to help pass the evening hours.

As you walked in from the highway to the BOQ there was a prominent sign that I observed every trip. This sign was purported to identify the current level of hurricane activity. It always indicated that we were in the midst of the worst hurricane in years. This information was painted on the wall under the heading of "Hurricane Condition," but should have been covered by several removable signs detailing varying degrees of severity that hung on two large nails. Of course. some years before. a hurricane had undoubtedly blown all of these lesser-degree signs to the four winds. literally. which made it a kind of automatic hurricane-measuring device. and they had never been replaced.

I worked with two different groups at the Key West Naval Air Station—VP-1 and ZX-11. The VP-1 Squadron was an experimental squadron where things like the ASQ-8 were tested in aircraft and the results passed along to the rest of the Navy. They flew their engineering ASQ-8 in a PBY *Catalina*. a durable but ancient twin-engine amphibious aircraft designed by Consolidated in 1933. This particular PBY. in which I made numerous flights. had been restricted to ground take-offs and landings because of some structural modifications. so I never got to make flights off water at Key West. But just flying in the PBY was a thrill. Cruising at ninety knots on a good day. I could stand in the rear with the bubble canopy

on the left side wide open, the wind in my hair, and have a wonderful view of the world.

The PBY also served another important purpose. I was told on good authority that the wheel wells could carry thirty cases of rum when returning from the Guantanamo Bay Naval Base on Cuba.

I was privileged to fly with Commander Lingle, the boss of VP-1. He loved to fly and his nickname was "Shot" Lingle. This, it turned out, was the diminutive for "Hotshot," as in "Hotshot Pilot." He was good. On one flight with "Shot" at the controls, we began our takeoff for a calibration run on the ASQ-8, and just as the PBY broke ground, I was sprayed with hydraulic fluid as I sat at the navigator's station. The hydraulic cylinders that boosted the rudder and elevator movements were fastened on the bulkhead a foot or two from where I sat, and a seal failure gave me a surprise dose of the oil. Shot was duly informed of this failure and shut off the hydraulic pumps, which stopped the spray into the cabin. He then proceeded to fly the entire mission without the control boost, which was akin to driving a bulldozer without power steering. When we landed a couple of hours later he was wringing wet with sweat but unconcerned about the failure. I never did get the spots out of my shirt.

Shot's reputation as a skilled pilot left him open to a little good-natured nitpicking by his flight crew. He told me the story of a flight where he had some time to kill in the air while waiting to rendezvous with a submarine for tests, and decided to practice single-engine flying. This is a good thing to know how to do well in case of an engine failure. He called the Flight Engineer, who sits all by himself in the wing pylon between the engines, on the intercom and told him to shut off and feather the port engine. He received a "Yes, sir, skipper!" in his earphones and the port engine was promptly shut down and its propeller feathered. Shot then honed his flying skills by making turns in both directions and maintaining altitude on just the starboard engine.

Satisfied that he could fly the plane just fine on that engine, he told the Flight Engineer, "Okay, that was fine. Now let's do it with the starboard engine shut down and feathered."

"Yes, sir, skipper!" answered the Flight Engineer who then shut off the starboard engine without restarting the dead port engine.

Shot told me it suddenly got very quiet up in the air with just the noise of whistling wind.

"Goddammit, what I meant was, restart the port engine and *then* shut off the starboard engine...!"

"Yes, sir, skipper!" replied the FE, I'm sure with a smile on his face. Gotcha!

My second favorite story from VP-1 was a much-heralded ground demonstration of the latest droppable tip-tanks on the Navy's jet trainer, the T-2. It was the same as the Lockheed F-80 *Shooting Star* as I recall. It was decided to demonstrate this gee-whiz device to interested parties by positioning a mattress-lined cradle under the wing-tip-mounted torpedo-shaped tank. It would only drop a foot or two into the mattresses and would not be damaged. The crowd was assembled and a maintenance mechanic dispatched to enter the cockpit and drop the tank for the spectators. Of course, it was the other tank that fell to the concrete and was damaged beyond repair. while the one positioned over the mattresses never moved an inch.

My favorite PBY flight with Commander "Shot" Lingle took us out for the usual two or three hours of passes at various altitudes over a surfaced submarine and recording the magnetic signals. We finished our chores for the day and I had moved up forward to watch Shot fly the plane which held a never-tiring fascination for me. He saw me and shouted. "Go back to the blister and I'll show you how we *really* hunt submarines!"

I went back to the open bubble canopy on the left side of the PBY and held on tight as Shot wheeled the lumbering patrol plane in a steep left bank and downward spiral toward the sub. We passed over the bow of the submarine at maybe fifty feet and full throttle. banked to the left just enough to give me a splendid view. A sailor in the conning tower of the sub who had been watching us with his binoculars, put them down and looked me right in the eye. his mouth dropping open, as we roared by. It gave me a great feeling of superiority.

One memorable trip brought several of the TI managers down to Key West with me for a meeting with the Squadrons to discuss ASQ-8 field service. Bob Olson. Chief Engineer of the Apparatus Division, being of sufficiently high rank and then some.

had rented a car for us. What a treat! And besides that, it was a red Plymouth convertible, which was the only rent car they had. But it seemed to me that a visitor to Key West would look just perfect in a red convertible.

Bob soon found a terrible flaw in the rent car. When the steering wheel was turned even a reasonable amount, like turning a corner, the horn would honk. For Bob, who was the epitome of the polite and courteous gentleman, it was terminally embarrassing in downtown Key West to honk at the pedestrians every time he went around a corner. He became adept at shrugging his shoulders and holding his palms helplessly up in the air.

The other half of my field service in Key West was devoted to the blimps of   ZX-11, the Navy's experimental lighter-than-air squadron. No big hangers here, the blimps were parked out on the tarmac attached to portable docking masts. If any scary weather was in the forecast, the blimps gassed up, so to speak, and high-tailed it out of there.

The usual ASQ-8 test runs were combined with what were euphemistically called "navigation training flights, but were really an excuse to take a nice leisurely blimp ride over the lunch hour. This allowed the clever sailors of ZX-11 to load up the blimp with good food and cook a better-than-average hot meal aloft. These K ships had everything, including a complete galley for meal preparation.

Properly stocked, we would take off ("Up ship!") and head almost due west out of Key West. The navigation goal, which could almost be seen with the naked eye if you got a little altitude, was the Dry Tortugas. This cluster of keys is famous for Fort Jefferson, built during the civil war to protect the most god-forsaken piece of real estate in the United States. It lies at the end of the string of keys off the tip of Florida. The fort was also used to incarcerate Doctor Mudd, convicted of patching up John Wilkes Boothe after he assassinated Lincoln. Fort Jefferson was and is a large and imposing octagonal structure of red brick. How they ever built it on the tiny key I'll never know, but they did and it's a minor marvel. It's also a Florida State park.

The blimp, having successfully found Fort Jefferson by dint of skillful navigation, would circle it and the usual shrimp boats, and

20

then head back to Key West. At this time, the first of the hot lunches would be ready to serve.

On this particular flight, the hamburger steaks with grilled onions smelled excellent and I had worked up a considerable appetite, having been forced to run the ASQ-8 for ten or fifteen minutes on the flight out. As a guest, I was politely served my meal before the captain, and I carried it back to the comfort of the rear seats to eat. Just as I had finished the last delicious bite, the captain came running back to where I was wiping my mouth and belching, and shouted, "Don't anybody eat this hamburger! It's spoiled!" Well, fine. I just finished about a pound of it.

I sat quietly in my seat for the rest of the flight and imagined how sick I was going to be. That thought plus the usual rolling and pitching of the blimp had turned me a bilious shade of green by the time we touched down at the base. Boy, was I glad to get back on the ground. But the skipper was wrong. Either that or I have a truly cast-iron stomach.

Key West was funky in the extreme in the early 1950s, and untouched by "modern" civilization. It was at the end of US 1, the Overseas Highway, from Miami, built on the ruins of the Flagler railroad. Quaint hardly described it. Most of the commercial structures, such as the Key West Cigar Manufactory, were frame buildings and either never painted or not painted in a long time. The buildings were a wonderful weathered gray color. The big downtown hotel, the LaConcha, was brick and on Duval Street, the center of all good and socially acceptable entertainment. Since everything in Key West was within walking distance of the Sub Base, I even ate there a time or two, discovering key lime pie in the process.

But the most dreadful place to stay in Key West was a three-story ramshackle frame hotel, using the term loosely. Whorehouse would be a more accurate term. And as it happened, I got to stay there two nights, courtesy of the Submarine Base commander, whoever he was.

I walked into the Sub Base BOQ, suitcase in hand, to stay a few nights on yet another trip to the Naval Air Station, and was told that civilians could no longer stay there. Oh, great. Now they tell me. Seems that the base commander thought that the civilians were taking advantage of the Navy by only paying a dollar a night for a room. He was probably right, but the timing of his discovery was poor. So, suitcase back in hand, I trudged out past the Hurricane Warning sign and headed for the La Concha hotel downtown. Since I was not the first civilian to be booted out of the BOQ, the La Concha was full. Tough luck. And of course the only other identifiable hotel in town was the frame whorehouse. It wasn't full, at least not if I wanted to pay the full rate and stay all night.

My room was on the second floor and it shared a bath with the adjoining room. The baths had been built at some later date and were tacked on the outside of the building with wooden struts propping them up from the ground. They were narrow and long enough to straddle two rooms. They were the pits and fitted the hotel décor perfectly. My iron-frame bed had a set of steel springs topped with a mattress and covered with ratty linens. This resided under a single bare light bulb that dangled in the center of the room.

After a poor night's sleep. what with the footsteps, bed-squeaking and door-slamming all over the hotel. I got up to get ready to go to work. I was doing okay until it was time to plug in my electric razor and shave. There was no outlet in either the bathroom or the bedroom. I was either the first person to stay overnight or the first with an electric razor.

I pulled on the rest of my clothes and went out to the street to see what I could find. There was a drugstore nearby and I bought a screw-in plug adapter for the light bulb socket in my room. I finally shaved standing on the bed in the dark.

But for every ramshackle whorehouse there was a wonderful place that I stayed while servicing ASQ-8s. The finest was a nameless hotel in Honolulu on Waikiki Beach. next to the Royal Hawaiian. where I spent a glorious week. A converted estate with only a dozen rooms. it was exquisite. Breakfast was served each morning on the lawn just up from the private beach. Here was where I learned about papayas. It also made up for the dollar a night I'd been paying at the BOQs.

My job in ASQ-8 field service allowed me to fly in most of the Navy patrol aircraft that were in service at the time. After I was married. TI carried a special life insurance policy that would pay some amount of cash to Mary Ruth in case I went down with the ASQ-8. I was insured while flying in unlicensed. experimental, and military aircraft. That about covered it. I flew in lots of Lockheed P2Vs of various models, a PBM out of the San Diego Naval Air Station, and the prototype of the Martin P5M at the Martin factory in Baltimore.

Probably the most uncomfortable ASQ-8 check-flight I ever had was out of the Oakland Naval Air Station in an AF. The AF was built by Grumman and used in the "hunter-killer team" tactics. One of the pair of airplanes would be equipped with the ASQ-8 and radar as the hunter; the other would have a group of weapons and act as the killer.

The AF was a really big single-engine tail-dragger and an impressive airplane. It had one huge radial engine on the front end with the biggest propeller in the world. The whole thing seemed impossibly large for a single-engine airplane. The "hunter" version I flew in carried a crew of three, with the pilot and navigator/gunner on the top row with a canopy to see out. Then there was the MAD and radar operator down below in the belly, with no place to see out. That, of course, was where I rode.

After pulling on my borrowed genuine Navy flight coveralls and donning a helmet and parachute, I waddled out to the flight line and met the crew I was to fly with. To simplify getting into my seat on the plane, I was picked up bodily and inserted into a seated-person-shaped space in the bottom half of the aircraft. A minute's instruction on working the intercom and how to operate the door latch in an emergency, and I was qualified to fly. Just before they shut the door on my coffin, the pilot stuck his head in and said, "Oh, yeah. If I have a problem and we have to bail out, I'll ring a bell." He pointed to something vaguely over my head. "The bell is cranked by hand in case we don't have any electricity in the plane, and if you hear it, just pop this door open and crawl out. You're low enough that you'll go under the stabilizer, so don't worry. And oh, yeah, don't forget to pull that metal ring there after you get clear. That opens your parachute."

In the high-tech world of Navy aircraft, the escape warning bell is hand cranked? Whatever. If I heard a bell, I would be outta there.

The flight was claustrophobic at best. There was a tiny window in the door at my left elbow but too high to look out of.

Equipment was in every conceivable nook and cranny, leaving just enough space for a seated person. With the seat belt and shoulder harnesses cinched up, you couldn't move anyhow. The MAD gear worked fine, but whoever has to sit in that seat for a living deserves every penny he or she makes. Bad way to fly.

The navigator on the flight told me later about a flight he had in the MAD station where I had ridden. Barreling down the runway at full power on takeoff, the engine suddenly quit cold. They continued rolling down the runway at a high rate of speed and the engine just as suddenly roared back to life and they continued the takeoff. He, being unable to see what was going on, got the fright of his life but didn't feel free to ask the pilot about it. They flew the mission without any more problems, and after they had landed and were walking back to the hangar, he said he had to find out, and asked the pilot about the take-off problem.

The pilot told him, "Oh, nothing serious really. About halfway down the runway on the takeoff, some loose papers began blowing around the cockpit. I made a grab for them and accidentally hit the magneto switch and shut the engine off. I stuffed the papers under my leg and turned the mags back on, the engine caught, and we just kept on going." I'm really glad that didn't happen when I was flying.

In addition to working with every Naval Air Station in the US and Hawaii, I worked with the aircraft companies themselves. I flew in the prototype of the Martin P5M out of Baltimore and got to observe first-hand the magnetic effects of firing the dual 20-mm tail turret guns. Yes, firing the steel gun barrels when they're lined up with the earth's magnetic field causes them to become slightly magnetized, and yes, I thought the airplane was going to shake apart while the guns were being fired. But taking off and landing on the ocean was quite different, and I'm told much more difficult than on a runway.

I spent more time at the Lockheed plant in Burbank, California, than with any other aircraft manufacturer. They were building the P2V *Neptune* patrol bomber and equipping some models with a "stinger" on the tail to carry the MAD detector head. This long molded fiberglass tailcone carried the rear contours of the fuselage out an additional seventeen feet to move the detector as far from the plane and its magnetic disturbances as possible. It was distinctive looking in flight.

Most of my work at Lockheed was done in their "MAD Shack" out near the edge of the main runway at the Burbank airport that Lockheed shared. Lots of interesting things occurred here, like the emergency landing of a *Navion* four-place private plane right in front of us. The MAD Shack was equipped with a tower-frequency radio so we were aware of what was about to happen. We all went outside and lined up to watch the action. The Navion, which had a complete farm gate—pipe frame, fencing and all—wrapped around its nose wheel, landed very nicely, keeping the nose of the plane as high as possible until it lost speed. The tire and all the debris finally dropped down onto the runway in a shower of sparks followed by the bang of the tire blowing out. The plane swerved just a bit but quickly came to a halt in the center of the runway. The canopy cranked open and, one at a time, four ladies emerged from the plane and walked away without a backward glance. Nice job.

Several times I was thrilled to see the monster Lockheed *Constitution* fly. As I recall, this huge experimental transport plane was built for the military but the new 36-cylinder "corncob" radial engines were not sufficiently reliable. But what a sight to see it lumbering down the runway and taking off. Once as it was climbing out after the takeoff, an engine suddenly begun boiling out black smoke. The Lockheed employee watching next to me said brightly, "Oh look! They've fired the JATO bottle!" JATO bottle my ass, the engine had just shelled itself. The *Constitution* swung ponderously around the field and landed with number two engine shut down and feathered, the cowling and wing covered with oil. Back to the hangar once again.

On several occasions Lockheed flew their ASQ-8-equipped P2V-5 to the Dallas Naval Air Station for us to use for engineering evaluations. This plane was owned by Lockheed, and would eventually be sold to the Navy, but at the time we worked on it, it was Lockheed's and came equipped with a Lockheed test pilot.

Everything you learned about test pilots from watching B movies appeared to be true. The day that I flew with the Lockheed pilot, he had been carried to the plane, still asleep after a night of partying, and stashed in a crewmember's bunk in the rear. We loaded our gear on board and when we were ready to go, they woke the pilot and he shuffled to the front, rubbing his eyes, and sat down in the left-hand seat.

After buckling up and flipping a lot of switches, he started both engines and, after some brief and unintelligible radio talk, began taxiing to the end of the runway. I had buckled in at the navigator's station and could watch the pilot's activities. As a pilot of a sort, I was always interested in flying machines. And the pilot was doing something that I didn't understand. As we taxied, he reversed the propeller pitch on one of the engines and revved up the other, and then did the same thing with the opposite engines. As we approached the end of the taxiway leading to the beginning of the runway, the pilot shoved both throttles to full power and stood on the left rudder pedal.

We accelerated rapidly and slewed across the taxiway and onto the runway with a roar, like a cornering racecar. By the time the plane was straightened out on the runway we were halfway to takeoff speed. I had somehow missed the magneto checks and few dozen other items that I thought a pilot would do before taking off. Then I realized what he was doing reversing the pitch of the props as he was taxiing—he was warming up the engines in a hurry. When we hit the runway at full throttle, the engines had only been running a minute or two, not nearly long enough to reach the proper operating temperature. Unless, of course, you'd been taxiing along with the propeller in reverse pitch.

We climbed above the scattered clouds and ran our magnetic checks, which involved putting the plane through a series of rolls, pitches and yaws. I had already learned not to eat before checking out ASQ-8s. The rolls and yaws aren't too bad, but the pitches will get you every time.

After I'd finished my part of the engineering work I moved up to the doorway behind the pilot and copilot and squatted down to watch them fly the big bird. It's a thrill to fly in a big powerful airplane such as the P2V. The jets can never match the gut-wrenching roar of those two 3,000 horsepower Wright-Cyclone radials.

When it was time to go back to the Dallas Naval Air Station, our pilot began a wide circle above the clouds. Then, without warning, he announced, "There's a hole!" and cranked the P2V over in a tight diving spiral. I watched the G-meter on the instrument panel reach almost two Gs, for which my squatting position was ill suited. I thought both legs were going to break in half at the ankles.

Our pilot had filed a visual flight plan if he filed anything at all, and needed to find a hole in the cloud layer big enough to fly through without instruments. And he did, even if it nearly crippled me. And then he greased the big bird onto the runway as slick as Teflon. The stories I'd heard about test pilots had not been exaggerated.

Although it may not sound like it, traveling and doing field service on the ASQ-8 was a minor part of my work for TI. Most of my time was spent in the Lemmon Avenue plant working with Manufacturing on problems, and otherwise trying to improve our product. Roger Webster was usually involved in some part of the design that I didn't understand, and it frequently was something to do with the sensitive magnetic detector element itself. And every couple of weeks, Roger's steel leg would become magnetized to the point that he couldn't walk into the lab without making the needle of the chart recorders run off scale. Then we'd lead Roger over to a chair and drag out the degausser, a coil of heavy wire big enough to put your, or more precisely Roger's, foot through. The degausser would be plugged in, and with an ominous hum, Roger's foot and leg would be run into the coil and then slowly back out again. Roger was now degaussed and once again welcome in our lab.

I walked into Roger's office one afternoon and was puzzled by a strange contraption hanging from his ceiling. It was a small cylindrical frame made of copper wires soldered together, with wire cones on each end. The whole thing was maybe six inches long and I could not imagine what it was for. I asked Roger, and he replied, "It's a battery eliminator."

I found that hard to believe, as a "battery eliminator" was generally a power supply of some sort. I asked him to elaborate and he did.

"Several of us were arguing about whether or not there's a magnetic field in the center of a current-carrying wire and nobody really knew. We decided to perform an experiment and see for ourselves." He reached up and took the strange device down and handed it to me to see. "We made up this hollow 'wire' out of bus-wire and put a little pocket compass in the middle of it."

Roger continued, "Then I took it over and touched the ends of it to the terminals of the car battery we had in the lab with everybody looking over my shoulder. The damn thing welded itself to the battery terminals and began getting really hot and I let go and told everybody to run."

"What happened?" I asked.

"The battery exploded. That's why I said it was a battery eliminator. It sure as hell eliminated the battery."

It was May 1954, and the time for our June 1 Semiannual Performance Review was rapidly approaching. However, I feared not. My work had gone well the past six months, and in fact my entire ASQ-8 career had been pretty good to both TI and me. With a few minor exceptions, I had caused no damage and most of the things I had designed worked. Not all, but most. And, it occurred to me, I had been an Assistant Engineer since December of 1951, far too long.

My Performance Review score sheet was just swell. I ranked at or near the top in all categories. But I was still a Job Grade 10 Assistant Engineer. I inquired politely why I didn't make Grade 12 Engineer after all these years, and was told that there weren't any "openings" in that job grade for this department. Already filled up, they were, with Grade 12 engineers. The thought occurred to me that I'd have to wait until someone croaked to get promoted to engineer. Not really very promising, although the possibility of putting out a contract on one of my friends passed through my mind. Rats.

But a few phone calls later and a trip to the Bowling Alley to talk to Cecil Dotson, and on June 1 I was transferred to Semiconductor Manufacturing. I was to be a Job Grade 11 Foreman, Semiconductor Assembly. So there, too. Take that! And it was the best career move of my life.

Chapter Three

The old bowling alley, two doors down the street from the TI Lemmon Avenue plant, was no longer a bowling alley. It was now a transistor manufacturing facility. After spending many happy evenings there swilling beer and missing seven-ten splits, it became my workplace on the first day of June, 1954. But the good news was that the name never changed. It was still the "Bowling Alley" even when it was full of crystal pullers and transistor assembly lines.

My job as foreman of a transistor production line was the standard break-in job for new semiconductor engineers. It was a terrific experience and the hands-on and practical aspects of actually trying to build the little boogers were invaluable. I was now the boss of four nice ladies, all of who knew a lot more about building transistors than I did. We were building the Texas Instruments 300 Series PNP alloy-junction transistors in the oval solder-seal package. Neat stuff. They came in three different models: the Type 300 with the lowest "gain" or performance, the 301 with an intermediate range of gain, and the 302 which was the hot stuff with a "beta" or current gain of 50 and above. Wow! Something for everybody! Those that met these specifications sold for ten to fifteen dollars each. Those that didn't went into the reject bucket.

Before long, the ladies had trained me to do all the operations necessary to build the transistors, although they did better with their delicate touch on certain of these process steps than I did. For example, holding a piece of microscopic wire with a pair of precision tweezers just so against the indium dot on the side of the germanium chip and spot-welding the other end to a header post while peering through a microscope. I never did get very good at that. But we all took turns at all the jobs, and it was hard to tell who was actually running the production line. We made a pretty fine team if I do say so, and on a good day we could crank out a hundred transistors, although sometimes I had to come back at night to finish canning them.

My other official Foreman duty was to "manage," using the term loosely, D. D. "Mac" McBride, sole producer of TI point-contact transistors. The point-contact transistor was the very earliest

29

type and the first transistor product put into production at TI. It was also the first to become obsolete. McBride would build a batch of thirty units every two days. test them. and put the good ones in stock.

Then someone noticed that no point contacts had been sold in months. My short and easy career of being Line Foreman of the point contact transistor assembly line (that's Mac I'm talking about) came to an end. McBride was a highly skilled and talented technician and was in great demand throughout the transistor plant and after leaving my so-called supervision. never looked back.

Jim Lineback. the original Process Engineer at TI if not the world. worked with me on the alloy-junction production processes to improve the yield of good devices and streamline the assembly. He was a great guy and I learned a ton from him. Later. when I became a process engineer. it was his basic training that carried me through.

Jim. like every other living person at TI. had to take the Work Simplification training course. Don't get me wrong. it was a great course and the remnants of its teaching stay with me yet. I enjoyed it although it seemed the classes were always at the most inconvenient times. Nevertheless. Lineback achieved a fame of sorts in the Work Simplification field by his motion analysis homework. The homework problem was to pick some activity that you do on a regular basis and analyze each and every motion required to do it. A chart was made using symbols for the types of motions. and then, of course. the idea was to study it and simplify it to save you untold hours for the remainder of your life. Or something. Anyhow. without going into any more detail. Jim picked the function that toilet paper is used for. His treatise was a smash hit with the Work Simp gang and was used as a good example of creative problem solving. I have forgotten exactly how much time Jim saved with this analysis.

On the other hand. Bob Brons   in the Transistor Production Planning group. had an even more cr   .e solution to his most pressing problem. In the introductory class session he learned the first rule of Work Simplification: study the problem and see if it can be eliminated entirely. Bob studied   biggest problem. which was interrupting his work to go to Work Simplification. and so he never went back. It worked!

Jim Lineback and I got along famously and I had great respect for his transistor production knowledge. But one day I was in a bind on my production line. Someone had carried off my one and only hotplate and I needed it to boil a batch of in-process transistors in alcohol. That's what the Lineback specification said: "Boil the transistors in a 100mL beaker half full of clean ethanol by placing them in a pith-wood block on a ceramic base and submerging the transistor elements in the liquid. Boil vigorously for five minutes."

As I wandered around vainly looking for my hotplate and carrying my beaker full of alcohol and transistors I spied an unattended vacuum pump and bell jar. Aha! You could probably have seen the light bulb light up over my head. I stuck my beaker of transistors under the bell jar and turned on the pump. In a few seconds the pressure inside dropped and the alcohol began boiling merrily, which looked about the same as vigorously, and I started my stopwatch. Just then Jim Lineback walked up and politely asked me what the hell I was doing.

"Boiling the transistors in alcohol like your spec says. Four and a half minutes to go." I replied, glancing at my stopwatch.

"Damn it! Not at *room temperature!*"

"Well, the spec doesn't say one way or the other..." He made me rewrite the specs and promise to do better. I rewrote the specs.

On an otherwise fine afternoon in the Bowling Alley the transistor production was interrupted by a loud, sharp explosion that I later learned to associate with hydrogen. This was followed by a tinkle of broken glass. A split-second later, the room was filled with the screech of a hundred metal chairs being pushed back from workbenches and the sound of a lot of feet running for the back door. By the time I had taken this all in, there was no one left to supervise on my production line so I headed in the direction of the germanium crystal pullers.

The crystal pullers were ingenious phone booth-sized machines designed by Boyd Cornelison. TI's sagacious semiconductor chief scientist. The crystal puller heated a teacup-sized crucible of germanium to a red heat, or about 950 Celsius. and then dipped a small piece of special germanium called a "seed crystal" into the molten pool. As the crystal was spun slowly and simultaneously drawn slowly out of the melt, a single-crystal germanium ingot would "grow" on its end. And that's the stuff you needed to make transistors. Just to make it dangerous, the melting and pulling had to be done in a hydrogen atmosphere, and keeping air out of it was necessary to keep it from exploding and blowing the enclosing quartz shield to bits. Something had obviously gone wrong as bits of the quartz shield were scattered around the puller.

Boyd was standing there, safety glasses slightly askew, cussing the machine up a storm. No one was hurt in the explosion because of the plastic safety panels outside the quartz tube, but Boyd was pissed. He'd ruined a perfectly good crystal by flipping the wrong toggle switch on the control panel. While peering intently at the conformation of the growing crystal he had felt his way over to the switch panel to turn off the spin motor and had inadvertently lowered the crucible with the wrong switch. This broke the seal and opened up the chamber full of hydrogen and red-hot germanium to the air with predictable results. It blew the hell up.

I walked over to the rear door to see if I could locate my production team and was amazed to see the entire transistor manufacturing force at the far end of the Johnson grass field behind the Bowling Alley. The fire door had been ripped off its hinges and lay flat on the ground with footprints from one end to the other. I never did find out what poor soul got to the door first, but it must have hurt.

Boyd Cornelison was the most creative person I ever met. Working with him, which I had the great privilege to do later, was an engineering inspiration. I just thought I knew how to do a few things. Boyd could do *anything*. And even better, he could make it work. One incident that I observed occurred when one of Boyd's engineers was unable to grow a new type of transistor crystal called "rate grown." This engineer was using the latest model crystal puller whose crucible temperature and crystal pull rate were controlled by two disk-shaped cams. The contours of the outside edges of these disks told the machine when to raise or lower the temperature of the molten germanium and to change the rate at which the crystal was pulled from the melt. A couple of months passed as the engineer cut cam after cam and tried everything he could think of to grow this new type of crystal. Nothing worked and Boyd finally lost his patience, which was in short supply at the best of times.

I saw a scowling Boyd carry a new pair of cam blanks back to our little machine shop. He sat down at a bench and with the aid of a steel pocket scale sketched some profiles on them, got up, walked over to the band-filer and personally filed the cams to his contours. He then went back out and installed them on the crystal puller, cranked it up, and soon had grown the first successful rate-grown crystal at TI.

Boyd had another interesting side. I discovered this shortly after I had met him, when I was attracted by a series of smashing noises coming out of his office. I gingerly peered around the corner as Boyd was putting the pieces of a destroyed telephone in a cardboard box. He saw me and said pleasantly, "Ed, would you mind taking this back to the maintenance people for me? It quit working." And it certainly had. Any equipment that malfunctioned within the reach of Boyd was in a hazard zone.

Similarly, a year or two later, Boyd bumped his head for the umpteenth time on a poorly placed hydrogen regulator on the wall in one of the labs. He didn't say a word, I'm told, but reached up and bodily ripped the regulator from the wall and twisted it loose from the hydrogen plumbing. He then threw it on the floor and continued on this way, hydrogen spewing from the broken pipe. And sure enough, he never bumped his head again.

As the production lines got more and more crowded in the Bowling Alley it became obvious that a major change was needed. The impetus for the change came from the oppressive temperatures in the summer from overcrowding and inadequate air conditioning. But it was for the effect of this heat on the transistors, not the sweltering employees.

Jim Lineback showed his bosses that we were failing good transistors because of the temperature. Hot transistors had higher leakage currents than cool ones, and the tests were specified at a cool temperature, like 25 Celsius. We had been throwing away good transistors because the Bowling Alley had been reaching 90 F. or 32 C in the afternoons. Good grief! The employees being hot are one thing, but transistors quite another! Let's move the transistor production over to the big, cool, Lemmon Avenue plant! And so they did in the late summer and fall of 1954.

Meanwhile, another group at TI was quietly making history. The development of the first transistor radio had begun. Pat Haggerty, far-seeing prime mover of Texas Instruments, had correctly anticipated that the time had come, technically speaking, to build an all-transistor radio to replace the vacuum tube sets that were on the market. It could be the first consumer product for the up-and-coming transistor. But the transistors at this time, May 1954, were just barely good enough to build a radio. Just barely, maybe. So he got Paul Davis, electrical engineer, who in the past had designed radios for Watterson Radio Company, to pull a team together and build a breadboard model, i.e., one that functioned but didn't look pretty. Fortunately, Paul later wrote up this once-in-a-lifetime experience in a piece he called "The Breakthrough Breadboard"© from whence these details are taken.

When Pat Haggerty said he wanted a working transistor radio, he didn't mean next year or next month. He meant next week. So Paul Davis and his team, consisting of Roger Webster, Ed Jackson and Mark Campbell, started with no designs, no specifications, no coils or transformers suitable for transistor use, and no transistors that had been used at these frequencies. Other than that, the prospects looked good. Kind of like a snowball in the Bowling Alley.

Pat had given Paul the challenge late on Friday, May 21, 1954. On Tuesday afternoon, the 25th of May, Paul and his team carried four pairs of blood-shot eyes and the world's first working transistor radio up to Pat's office on the second floor of the Lemmon Avenue building.

As Mr. Haggerty was fond of saying, the reward for a job well done was a harder job, and it didn't take long for Paul Davis and his team to be heavily rewarded. On the following Saturday morning, just about the time their eyeballs had gotten back to normal, Pat dropped by Paul's office and asked them to redesign and build their new radio circuit in an empty Emerson vacuum-tube portable radio case. But he didn't need it until Monday evening when he was flying to see a customer. This involved redesigning and building smaller transformers and coils and other parts that weren't commercially available. They quickly enlisted the aid of Harry Waugh to design and build a tiny on-off switch and to fit up a chassis for the new parts. Of course, they made this utterly impossible schedule because Pat Haggerty asked them to. Few people could inspire engineers like Pat.

This lead to an agreement with I.D.E.A. Corporation in Indianapolis for the joint design and production of the Regency Model TR-1 transistor radio. Jim Nygaard had just graduated from

Texas A&M and was added to the TI team, and Dick Koch from
I.D.E.A. was their chief engineer. The rest, as they say, is history.
After technical difficulties that I will not bore you with but I assure
you were monumental, the Regency radio hit the Christmas '54
market with a monster splash. It was a big hit at $49.95, and
eventually over 100,000 were sold. And each one had four TI
transistors in it.

I was working in Jim Nygaard's group in the spring of the
following year, 1955, and he engineered the neat deal of the
decade—he got each of us, about fifteen people altogether, a set of
Regency TR-1 parts, less only the transistors and the resistors. We
even got to choose the color of the case. Then he acquired the
necessary transistors and resistors for us and we built our own
Regency radios. I still have mine. Bless you, Jim, it's a treasure.

By 1955 the quality of transistors had improved to the point
that transistor radios could be "cookbook" designed, particularly if
you used Roger Webster's famous "How To Design IF Circuits For
Transistor Radios" technical paper. During the summer vacation of
1955 I had the delightful task of not taking a vacation and instead
designing and building a larger portable radio for the Warwick
Company. Warwick built radios for other labels, such as Sears, and
had asked TI to design one for them that would meet certain
specifications.

After waving goodbye to my friends as they left on the plant-
shutdown vacation, I sat down at my desk in the now-quiet Lemmon
Avenue building with a copy of the Warwick specs in one hand and
Roger's "How To" in the other. It was a rewarding two weeks of
work. I designed and wound all the coils and transformers for the
radio, laid out and built the chassis for mounting the parts, and then
assembled and wired it. Lastly, I band sawed up a block of sugar
pine to make the cigar-box sized case. I got Virginia Miles, Mark
Shepherd's secretary, to pick out the color to paint it. I could do
radios but I couldn't do décor.

Sometime along in this early radio era, a call came into TI
from the downtown Dallas Chrysler automobile dealer. The caller
was shunted around until he ended up with one of our transistor
engineering managers. The dealer was desperate and as a last resort

35

had called TI. He was setting up for an automobile show on the weekend, and the center-piece of his exhibit was a Chrysler "concept car" that the factory had loaned him. And the center-piece of this concept car was the world's first transistor automobile radio built by Philco. But, as the now hyperventilating dealer explained, it had just quit working as they were playing with it. Could TI possibly help?

The manager, more clever than most, replied, "Sure! Bring it out and we'll see if we can fix it for you!" The dealer couldn't see him rubbing his hands in glee, at the opportunity to get an inside look at one of our competitor's top-secret radios.

A good old-fashioned TI fire drill was generated on the spot. A half a dozen of us were conscripted into two teams: one team to analyze the failure and attempt to repair the radio, and the other to reverse engineer it and come up with a schematic diagram of the circuits. I was on the latter team. And, oh yeah, don't bother going home for dinner, and plan to spend the night.

The dealer brought us the defunct radio personally, along with fervent pleas, and it was taken straight back to our waiting lab. We crowded around it and admired the work of our fierce competitor, Philco. It was a gorgeous radio. We promptly skinned it down to the chassis and were even more struck by the internal workings. It was a very complex and advanced radio, much more so than the battery-powered portables that we had been working on. This was going to be an interesting night.

Our two instant teams shared the radio the best we could, with the failure analysis guys getting first dibs with their instruments. We of the schematic-stealing team began a rough layout by counting the transistors, diodes and RF, IF and audio transformers. It didn't take Henry V. Stewart III, better known as Hank Stewart, long to find a dead IF amplifier transistor. It was one of the dreaded Philco "surface barrier" devices, a patented PNP transistor that outran anything we had. It had gain out the kazoo, as Jim Lineback used to say. We had nothing in our bag of products that could match it, or replace it in this sick radio for that matter. Hmmm. What to do.

"Wait a minute! I've got an idea," interrupted one of the transistor product engineers. "I was running a batch of experimental PNP units last week and they varied all over the map, but there were a couple of sports with really high gain. Let me go look in my desk drawer." And so it came to pass, a fluke transistor, cunningly resymbolized as a standard TI low-gain part to some day give the Philco radio engineers a chuckle, was installed and the experimental radio came to life. The radio was a marvel with automatic signal-seeking and sensitivity beyond anything we'd seen.

As dawn broke we tidied up the lifted schematic of the Philco radio. It was really sophisticated and well-engineered. It was a very nice piece of work. We admired it between yawns, and then put the covers back on the radio. It had been an interesting night.

I have in my possession a sheaf of transistor radio schematics and specifications, unfortunately not dated, but most likely from the 1955-1956 era. TI engineers had designed these radios and the information was available, along with the proper kit of transistors of course, to anyone wanting to go into the radio business. These eight radios range from a little four-transistor job with 35 milliwatts (that's 35 thousandths of a watt) of audio power up to a "big" six-transistor one that sported a push-pull output delivering an ear-splitting one-third of a watt of sound. By contrast, the Regency TR-1 had a maximum audio output power of 18 milliwatts. In today's world of 500-watt home stereos, that sounds impossibly weak but it could fill a room with sound, believe it or not. Well, a small room without a window air conditioner.

The production of transistors for radios ramped up at an impossible rate and Jim Nygaard's transistor test equipment group that I was in worked night and day in an attempt to keep up with the demands of production. The problem that I was assigned was finding a better way to test the popular 2N185 transistor that was used in the audio circuits of radios. It had a particular problem because it was used in pairs, as a "push-pull" output. Can't you just imagine one transistor pushing while the other one pulls? In any case, it required the two transistors to be very similar in electrical characteristics, like a "matched pair." Go look on your closet floor and you'll see what I mean by matched pairs. Since at that time, to quote good old Bob Chanslor, "We couldn't hardly make one alike," matching up our shotgun distribution of transistor variations was an almighty tiresome chore. Lots of ladies sat poking 2N185s in test sockets and then putting them in little piles on the tables. And then taking the piles and sorting them into smaller piles. Everybody had piles.

So I built a swell machine to match up 2N185s automatically. The operator would load a transistor in the test socket and push a "Go" button. Two seconds later, after a bunch of ratcheting sounds from stepping switches, one of 108 plastic drinking glasses would light up and the transistor would be tossed into it. It was a swell machine except for one slight oversight on my part. Only one person in the Lemmon Avenue plant could change a burnt-out light bulb in it, and that was Grace Weatherall, a lovely and petite secretary. Her slim hand alone could reach through the drinking glass support hole and unscrew the bulb. She was a very good sport about it, considering there were six machines, each with 108 bulbs, and she had a lot of typing to do.

The main difficulty in the design of this useful machine was deciding what kind of "bucket" to put the tested unit in. Seemed like having a light tell you where to put it would be a really good idea, but nothing rang my bell. Nothing until I once again went to the BBQ place in Richardson for lunch. And there, on my tray, along with a sliced beef sandwich, was a clear plastic drinking glass of just about the right size and shape. I carried my glass of water up to the cash register and said I wanted to buy it. After the confusion about water vs. the container was settled, I parted with two bits and the transistor collector was mine, conveniently containing the manufacturer's name on the bottom. Another high-tech hurdle was cleared.

The new Lemmon Avenue manufacturing facility was greatly expanded from the Bowling Alley as transistor production increased. Centered near one end of the large assembly room was the office of the big boss, Manufacturing Manager E. C. "Steve" Karnavas. Steve had been with the transistor group since its inception, and had joined Mark Shepherd, electrical engineer and boss, and Boyd Cornelison, physicist, as the resident mechanical engineer. His first beautiful little point-contact transistor-making machine is now in the Smithsonian.

Speaking of which, in the early 1990s, Steve let us all know that he was thrilled beyond belief that his point contact machine was now on display at the Smithsonian, in a special exhibit of transistors and technology. A friend of Steve's was the first to see it, and promptly phoned Steve in Dallas. "Steve, I hate to tell you this, but the display says the point-contact machine was built by Russell Karnavas. Is he any kin to you?"

What a rotten trick. You finally get something in the Smithsonian and they get your name wrong. A quick call by Steve fixed it, although when Steve and his family went to D. C. to see it in person, he could tell that a piece of paper with "Steve" had been pasted onto the exhibit to correct it. Whatever. That's a hell of a deal to make the Smithsonian, no matter how they spell your name. Good work, Steve.

I got to know Steve better one day when I walked past his office and saw him on the floor underneath his backup table carefully collecting scraps of paper. I knew there must be an interesting story, so I stopped and asked why the strange pastime. Steve got up off the floor and into his chair and spread out the collection of paper pieces on his backup table. He began matching them up like a cheap jigsaw puzzle. "Well," he said, "I just finished giving XXXX[1] her performance review, and after I finished she screamed obscenities at me in some foreign language and then tore her review up into little pieces and threw them on the floor. She does this every six months." He was making good progress in getting it together, and continued, "But I have to turn in everybody's review sheet, so I just tape all these pieces back together and turn it in with the rest."

"Gee! That's terrible!" I said. "She shouldn't do that! Why don't you get rid of her?"

"I can't," he shrugged. "She's too good a worker. And besides it doesn't take me long to put this back together. It's just a kind of ritual we have every six months."

Steve was also known for his creative approach to the lack of sufficient rest rooms. The sudden expansion had overtaxed the building's facilities and although more were scheduled for construction there was a period of overcrowding in the johns. And waiting outside till you could get in was not a viable option. If you needed to go to the restroom, you came by Steve's office and looked in the Ping-Pong ball bucket—blue for men and red for the ladies. If

---

[1] Not her real name.

there was a Ping-Pong ball of your color, you took it and went to the toilet. If not, you came back later when someone had returned one.

But one of Steve's continuing problems was that of the line personnel using the transistor burn-in ovens to heat their lunches. Burn-in ovens were used to artificially age transistors at high temperatures and weed out the weak ones. There were a lot of burn-in ovens in the final test area, and in addition to improving transistors, they worked really well to heat edibles. Too well, in fact. More than one can of beans or chili had exploded in the ovens, causing expensive and truly repulsive problems. Having a couple of thousand dollars worth of transistors coated with chili was no laughing matter, although I'll have to admit that I've had a chuckle or two after seeing the results.

Probably the strangest result of this employee bad habit was when a can of pork and beans blew its top as the oven door was being opened, the boiling contents narrowly missing the perpetrator. But across the room, a technician who was minding his own business and hard at work on the bench, suddenly yelled bloody murder and jumped up in the air. Later investigation found that a single really hot bean had sailed clear across the room and made a direct hit in his ear. Talk about a lucky shot! Threats of unemployment from Karnavas finally stopped the practice and made it safe once more to go near the burn-in area at lunchtime.

By this time I was working for Boyd Cornelison and building equipment for the production lines. Boyd taught me how to use the little oxy-hydrogen jeweler's torch and weld a microscopic bead on the end of a pair of platinum thermocouple wires. I became the de facto furnace thermocouple maker for the transistor group.

Boyd made that sort of tiny work look easy. He had come from Houston Technical Laboratories a few years before. TI bought them out and acquired, in addition to Boyd, a gravity meter business. Boyd had been the coinventor of the Worden gravity meter, the principle product of HTL.

The Worden gravity meter did just that—it measured the strength of the earth's gravity. The principle users of the Worden meter were oil exploration companies. Gravity variations could be used to find subterranean salt domes, which were likely places for trapped oil deposits.

The guts of the gravity meter were about a dime's worth of quartz made into a penny matchbox-sized collection of springs and torsion fibers. The finished instrument could tell the difference in the earth's gravity between the floor and the top of a desk. Boyd built these gravity-measuring elements with his handy-dandy oxy-hydrogen torch, a head-mounted magnifier, and a brightly-lit workbench topped with black Formica so he could see the hair-like quartz fibers. And, oh, yeah, a dime's worth of quartz rods. So welding the ends of a couple of pieces of ten thousandths diameter wire together wasn't much of a challenge for Boyd, but it was for me.

   Working with Boyd was one of the highlights of my career at TI. He was unfailingly pleasant to work with and was always doing something interesting. And he let me try interesting and strange things, some of which didn't work. I learned a lot.
   Probably my favorite lesson from Boyd was that there was always an easier way to do something, so sit down and think about problems before barreling into a solution. He gave me a living example one day after he'd asked me to can up a "stick" of ten transistors with a vacuum inside for a test he wanted to run. This, friends, was not a trivial task. At least it wasn't for me.
   After a few minutes of head-scratching on my part I decided to build some special transistor cans with a little piece of copper tubing sticking out the top of them. I would solder a tubed can on each transistor, and then, one at a time, connect the tubing to a vacuum pump. After it had pumped for a bit, I would crimp the tube really hard to seal it, cut it off and solder the end shut. This was assuming that I could really seal the tube with my crimp and not lose the vacuum before I soldered the end closed. I figured I could get it done by the next afternoon, easy.
   Boyd listened to my plan and thought about ten microseconds before he said, "Why don't you put a dab of solder on top of each can and then jab a little hole through it with a scribe? Then you could line up all the transistors in a vacuum bell jar under a piece of nichrome wire. After you've pumped the bell jar down, shoot some juice through the wire and get it hot enough to make the solder flow back and close up the holes all at once." Damn! Why didn't I think of that? Because I wasn't Boyd, that's why. Not much more than an

hour later I had canned up ten transistors in a vacuum and was on to something else equally interesting.

I knew I had arrived when one day, across the crowded TI cafeteria, Boyd stood up, yelled "Millis!" and gave me the finger along with a loud raspberry. Loud enough that everyone in the cafeteria turned to look at me. It was a proud moment. Boyd didn't give the finger *and* a raspberry to just anyone.

Late one night as I was working in the back of the transistor area, I heard an intermittent tinkling noise, as if something was falling out of the steel roof girders. I investigated and found Boyd around the corner in another lab, by himself, testing a large pile of transistors at a desktop electrical test box. He was sorting them into piles, but occasionally would throw one with great vigor into the rafters. He allowed as he was "roofing them," which was his proscribed fate for failed transistors.

Later, in similar circumstances, I heard him verbally sorting the tested transistors with, "That's a GSB. And another GSB. Oh, oh, there's an MSB. Um, that one's an ISB..." His three piles of transistors, the GSBs, the MSBs and the ISBs turned out to be Good SonofaBitches, Miserable SonofaBitches, and of course, Inbetween SonofaBitches. That terminology has been in use ever since, at least by those of us who were blessed to have worked with the most creative man at TI, Boyd Cornelison.

Typical of the things I was called upon to design, build and install was a timer system for the production people who etched transistors in "fume hoods." This nasty operation required wearing rubber gloves, rubber apron and a face shield, and gingerly holding a transistor by its little legs while you stuck its head under a stream of acid for a prescribed number of seconds. This was done in a fume hood, which was ventilated to keep the acid vapors away from the operator, but added to the difficulty. The stopwatches used to time this operation lasted a few months and then dissolved in a pile of rusty parts from the acid. I was told to find some better way.

I built a one-second timer and connected a half-dozen or so high-frequency buzzers to it, one in each fume hood. All the etch person had to do was to count seconds as soon as the device was in the acid and then take it out when the proper number of seconds had been reached. It was easier than the stopwatch since the operator didn't have to look up from the etching to check the time.

So, late one Saturday night I finished wiring up the hoods and turned the timing system on. It made a nice background "beep" every second that could be heard pretty much in the whole area. Perfect. And even better, I had one beeper left over.

It seemed appropriate that Steve Karnavas should be punished for my having to work late on a Saturday night, so I pried a piece of the metal trim from his office wall. I installed the last beeper inside the wall and snapped the trim piece back in place. Since I was careful to run the wiring on top of an existing electrical conduit, the installation was invisible to the eye. I went home happy.

Luck ran my way yet again. Steve came in Monday with a hangover and began immediately complaining about the beeping noise in his office. An unknowing colleague of mine, collared by Steve to help him find the source, was himself convinced that it was merely an unlucky reflection of the sound from the fume hoods, and he'd probably just have to get used to it. Steve continued to worry about it since it wasn't doing his headache any good.

Later in the afternoon I was passing by and saw Steve with a rolled-up piece of paper held up to ear, using it to listen to the latch hole in his door frame. He motioned for me to come over.

"Ed, the beeping is driving me crazy. Listen to this damn thing. It sounds like it's coming out of the door frame!"

I took the paper tube, solemnly put it to the hole and listened intently for a few
seconds. "I don't hear a thing." I said as I handed him back his tube and walked off.

But of course, he finally spotted the hidden wiring, pried the trim panel off of his wall, and promptly accused me of being the perp. How dare he.

Steve later became famous in a very, very small way by teaming up with his good friend Jack Kilby, inventor of the computer microchip and recipient of the Nobel Prize. They founded the Church of the Mechanically Inclined. Its primary object of worship is The Immaculate Contraption.

In the summer of 1955 I began my career as would-be inventor by submitting my first patent disclosure to the TI legal

43

department. It was for an alpha radiation detector and used the new TI Type 800 phototransistor. I thought it was pretty swell invention, since it used a material that would fluoresce in the dark red light spectrum when struck by alpha particles, and then transfer this light to the red-sensitive junction of the phototransistor. What could go wrong? I proceeded to build one and then tested it by borrowing a one-millicurie radiation source that just happened to be lying around one of the Apparatus Division labs. I unscrewed the radioactive "source" from its lead-block home and, holding the handling rod by the very end, waved it over my new invention. Absolutely nothing happened. I screwed it back in the lead block, disappointed.

A week or two later I happened to cross paths with Dr. Gordon Teal, Director of Research at TI, and a true gentleman, scientist and scholar. I gathered up my courage and introduced myself to him and, at his encouragement, poured out my puzzling problem of the failure of my invention. He never hesitated in very kindly suggesting to me that the total energy of the radiation striking my detector was very, very small, and likely not enough to cause any change in the operating point of the transistor. Of course! Silly moi! It was an excellent lesson in overlooking the fundamentals of a system, like the energy levels. I retrieved my patent disclosure and trashed it. Oh, well.

About the same time a second great idea popped forth unbidden from my occasionally fertile brain. This idea also used the Type 800 phototransistor, but modified in such a way as to make a motion sensor, otherwise known as a mechanical transducer. Another disclosure was written and optimistically submitted to Mack Mims, chief and only patent attorney at TI. This idea worked, although the commercial applications were approximately zero, not that this ever deterred me. I built an experimental unit and tried it at home as a phonograph pickup.

The practicality of the model was minimal since it required the phonograph operator to shine a flashlight down the end of the tone arm to make it play. It was nearly impossible to dance to the record music and operate the flashlight at the same time. But it worked.

The great day came in December of 1955, when Mack Mims called me to his office and shoved some papers in front of me. "We're getting ready to file your invention of the transistor transducer with the Patent Office, and I need you to sign the assignment form. This turns over the rights of your invention to Texas Instruments. Sign here, please."

I can tell you I was plenty excited! This was the way Edison started! Probably with a better invention, but the basic idea was the same. I read the brief form, which began, "For one dollar and other good and valuable considerations, I, Edwin Graham Millis, do hereby..." Wow! In addition to the thrill, I get a buck!

I turned to Mack and said, "Where's my dollar?"

Mack puffed on his pipe a time or two and then smiled and replied, "We don't actually give you a dollar for signing. That's just legal stuff."

For some reason it pissed me off big time. This great big rich company can't even supply me a token dollar on the greatest day of my life? I wanted something to frame and hang on the wall.

I said, "Well, Mack, then I don't actually sign this thing." I left without signing.

Several days later I was whining about the dollar thing to T.E. Smith, a man much older and wiser than I. T.E. also thought that was a pretty crappy deal, and probably not altogether legal if it should go to court. He said he'd take my case and we'd fight it to the TI Supreme Court, which at the time was J. Erik Jonsson, President of TI.

The following week I was called back into Mack's office and asked politely if I would please fill out a check request (on *my* Cost Center!) for a dollar, and he'd put it through for me. And then, would I please sign the form? Done and done. T.E. had won the battle for me, kind of. But it did break down the mythical dollar barrier. Soon the patent department, namely Mack Mims, had a bundle of nice fresh dollar bills and would hand out one to each assignment signer. Justice had been served for the TI inventor.

In 1955 I had moved into the Transistor Test Equipment group headed by Jim Nygaard. Jim, a recent graduate of Texas A&M, had worked summers as a co-op student and had been a friend of mine for several years. He was an extraordinary technical manager and my career at TI wove a convoluted path back and forth between Jim Nygaard's groups and everywhere else. He kept his people on the absolute cutting edge of technology and wasn't afraid to go to the top to get funding for a new and revolutionary idea. Jim was one of the important reasons that the TI semiconductor group grew and prospered. TI had better test equipment and better production equipment than anyone in the world. And Jim was the cause of a lot of it.

Jim also caused a lot of trouble. No one was safe from his twisted humor. Not even the janitor, Fred Brooks, who cleaned the halls and buffed the floors of the Lemmon Avenue plant. Jim took delight in waiting until Fred had turned the corner of the hall with his buffing machine, and jerking the plug out of the wall. Of course, when Fred came back to see what had happened, no one was there.

To bug Fred about the efficacy of his cleaning practices, Jim dug up a clump of Johnson grass and tucked it in a corner under his desk. When Fred came by sweeping the offices that evening, Nygaard motioned for him to come over. Jim bent over and pointed to the grass growing under his desk, and said, "Fred, what do you think we ought to do about that?"

Fred, never batting an eye, replied, "Keep it watered, I guess." Fred one, Nygaard zero.

Nygaard's test equipment group fit his spirit of adventure perfectly. Misfits is probably a little strong, but not far off the mark. And the leader of this opposition party was Earl "Mac" McDonald. Mac rarely missed an opportunity to goose Nygaard or cause some sort of uproar in the plant.

My very favorite McDonald episode occurred on a Saturday when Mac had really wanted to do something besides work at TI. But he had been "convinced" by Nygaard that it would be really good for TI and consequently his career if he finished building the transistor test set he was working on. Also, he was told, Mark Shepherd would be bringing a group of big brass from IBM on a tour through the plant Saturday and it would look good if someone like Mac were hard at work. Of course, a transistor contract with IBM would be the ultimate sale.

Saturday, on schedule, Mark Shepherd ushered a small group of coat-and-tied gentlemen into the lab where Mac was hard at work on his test set. Mark explained to the IBM visitors about the test equipment group, but Mac never looked up from his soldering iron. Mac then pushed back from his bench and walked over to a pencil sharpener on another desk to put a new point on his pencil. He couldn't quite reach the desk because of the chain on his ankle fastening him to his bench, but by stretching his leg out, he was able

46

to do his pencil. He went back to his bench, kicked the loose chain underneath, and began writing in his notebook, seemingly oblivious to the visitors.

Mac's combination desk and workbench, or "hutch" as they were called, was near mine, with its back to Jim Nygaard's office. Its tall top precluded Jim from seeing Mac as he worked, so when he was needed, Jim would just yell, "Mac!" In what became a standard sequence of events, there would be no reply. I could see Mac from my vantage point and he would totally ignore Jim's call.

"Mac! C'mere!" Still no response, but Mac, timing his motions to the split second, would slowly get out of his chair.

"Mac? Is Mac over there?" Jim would ask no one in particular. Jim would then walk out of his office toward Mac's hutch as Mac would quietly slip into his kneehole space and pull the chair in after himself.

"Damn! I though sure I'd seen Mac here a minute ago." Jim, puzzled, would walk back into his office as Mac quietly returned to his work position.

Mac was a good engineer and designed and built some of the most difficult transistor test sets. At one time he had a continuing problem with the electrical pickup of "hum" from the building wiring in the sensitive circuits he was designing. He asked Nygaard several times if they could get a good "earth ground" for our area, which would have fixed the problem, but for some reason Jim couldn't pull it off. Instead, he brought Mac a paper coffee cup full of dirt with a wire hanging out, labeled "Earth Ground."

"Here's your earth ground, Mac," said Nygaard. "Now quit bugging me."

Weeks later, Nygaard was at Mac's hutch discussing a test equipment problem and noticed that the circuit he was working on was wired to the dirt-filled coffee cup joke he had given him. "Very funny, Mac!" said Jim. Without a word, Mac reached up and took the cliplead from the cup's wire and the hum level on the oscilloscope jumped up to a high level. The damn thing really was an earth ground. Once again, Mac scored on Nygaard.

The problem of testing all the transistors TI was cranking out was beginning to be a major headache. The testing system was unchanged from the beginning of production several years before. Each separate test for a transistor required a test box and an operator to run it. The test box told the operator if the transistor was good or bad, or, occasionally, what range of goodness it was in. So there were a lot of test boxes on a lot of tables with a lot of operators passing transistors from one to the other. And we, in the test equipment group, were drowning in orders for more test boxes. Things were going to get worse before they got better.

Cleverly recognizing the problem that was beating us senseless, we kicked around some ideas for automating the testing. There had been an automatic test machine of limited use already built, but it was a circular "wheel" type of machine, with testing positions around the outside of it, and when you used up the 360 degrees, you couldn't put any more tests on it. It was a clever machine but it had a problem of not making reliable electrical contact with the transistor leads, and so fell into disfavor for throwing out good units because of the bad contacts.

But another problem actually caused the tester to be officially trashed. When a transistor passed all of the tests and was declared electrically good and salable, it was ejected from the machine by a puff of compressed air. The unit flew down a commode tank tube, which guided the unit clear of the wheel and then turned it downward into the "Good" bucket. Unfortunately, when it hit the elbow in the metal toilet tube to turn it downward, it was traveling about 200 miles per hour. The ricochet into the good bin would convert a fair percentage of good units into bad units in midair. We needed more good units, not less, so the wheel tester was abandoned.

We in the test equipment group began brainstorming about the test problem and developed an idea for a modular transistor test machine that would be built up like tinker toys—just keep adding on what you needed. It could have almost any number of testing stations and the same for the sorting stations. Each transistor would ride in its own little test block through testing stations until it came to a sorting station that checked the results of the tests and decided if it qualified for selection. If it did, the transistor would be pulled from the block and put into the bucket. A piece of punched paper tape would travel alongside the test blocks, keeping track of the passing and failing of each test. The sorting station would look at the array of holes and decide the transistor's fate.

We had pretty much worked out all the details of how we thought the system ought to be built, but Jim called one last meeting before putting together a proposal to present to management, a.k.a., the people with the money. We needed to think of a name for it. We argued for an hour with no particular progress until Earl "Mac" McDonald, ace troublemaker and test set builder, declared, "How about 'Centralized Automatic Tester?' We could call it the CAT Machine." We all liked that a lot and the name was adopted. It lasted a long, long time.

A CAT Machine proposal was duly generated, and I got to do little free-hand pen-and-ink drawings with my crow quill pen for the illustrations. It was submitted to Jim Nygaard's management and summarily rejected. Well, shit. But as it turned out in the long run, which is hard to see in the short run, not getting money to build this idea as we had envisioned it was good. It was much later that I realized we built better machines when we had little time and less money because it required us to simplify, simplify and simplify some more to save time and money, and to go with simple, fundamental things that we knew would work. But we were disappointed and put the development of the CAT Machine on the back burner and kept cranking out more and more hand test sets.

One hand test set in particular had been a real bitch, if you'll pardon the expression. Mac had been busting his buns night and day getting it built and the pressure from manufacturing was growing daily. Cecil Dotson, head of the transistor division, had personally gone to Jim Nygaard to try to get the delivery speeded up. Every hour or two, the line foreman from the test area would come by and ask Mac how it was coming. Late one afternoon Mac finally finished the set, and with a big sigh of relief told Nygaard it was ready to go to the line. Jim was now reluctant to simply give it them, what with all the trouble they'd been causing, so he asked Mac to help with a little presentation ceremony the next morning.

And the next morning the production line came to a screeching halt when the wail of a Scottish bagpipe rent the air. I was standing near McDonald when he turned it loose, and I had no idea how loud a set of pipes was indoors. "Rent the air" is not an idle simile. Maybe that's why they're played outdoors so much of the time.

49

Mac with his kilts and his pipes led the parade, his face the color of a ripe tomato from the exertion of playing some quaint Scottish air. The test set followed on a velvet cushion carried by Jim Nygaard, and our whole gang bringing up the rear. We traipsed into the production test area where the set was then ceremoniously transferred from the pillow to the tabletop. That demonstration pretty well stopped manufacturing from bugging us. They lost a lot of production that morning.

Our test equipment section was located on one side of a metal partition that separated us from a group of mad chemists developing a process for making silicon, which by this time was rapidly becoming the material of choice for transistors. This silicon was to be made from silicon tetrachloride, a "fuming liquid." Silicon tet, as it was called, was more or less readily available and could be turned into pure silicon by the magic of chemistry. It could be, after a lot of trial and error that is, and the errors ranked high in the trials. More than once I stood on my bench top and passed fire extinguishers over the wall to the brave residents of that lab.

Our room had a door that connected to the silicon manufacturing lab, but it was to be used strictly for emergencies. I guarantee nobody in our group would go through it into *that* lab, no matter what the emergency. We just kept the space clear on our side in case they wanted to come into our area in a hurry.

And one day there were some anxious sounds of a problem from over the partition, and a strangled-sounding highly-educated voice yelled, "Uh, say! Let's everybody get the hell out of here. It's a big spill." We all perked up and began to pay attention. Soon, a cloud of yellow vapors began seeping through the cracks around and under the emergency door, and more came wafting over the partition. Then somebody in our group said, "Uh, say! Let's everybody get the hell out of here, too!" And so we abandoned our ship of test sets and headed for clearer air.

Several hours later, after the clean up and ventilation had been completed, we returned to our work quarters. Every piece of bare steel or iron was now covered with rust—the drill press, our pliers and screwdrivers, even the door hinges. Later we found out that when fuming silicon tetrachloride gets mixed with moist air it generates hydrochloric acid fumes. If you ever want to rust anything, that's the stuff to use.

As we were looking over the damage, Johnny Ereckson, top technician in the test equipment group and the last one out the door during our abandon-office drill, said in amazement, "Look at my socks! They were blue when I put them on this morning and now they're red! No wait, they're just red above my shoe tops!" Sure enough, Johnny had been wearing litmus paper socks and didn't know it.

I segued back into designing transistor production equipment sometime in the summer of 1956. This work brought about my first issued patent, which was not the transistor transducer, but the "Electronically Controlled Bonding Machine," filed November 1, 1956. It was issued August 11, 1959, and the thrill of holding a copy of a patent with my name on it was great indeed. And of course, I had gotten another dollar.

During 1956, Jim Dixon, ace attorney and another good guy, had been added to the TI patent staff, instantly doubling its size. That summer I submitted a disclosure of my latest invention to him, the solar-powered flashlight. As my crude drawing showed, a person would don the pith helmet fitted with a solar energy panel on its top and wired to a hand-held flashlight-looking object. When the person stood in the noonday sun, the flashlight would light up. Jim presented it to the patent committee for consideration, and although I understood it got some favorable commercial possibility comments from Buddy Harris, it was tabled permanently. No dollar and no patent.

I must confess this idea was not totally original. Ed Jackson of the Research group had built a similar device except that the solar panel on the pith helmet was wired to a chin-strap-mounted tiny fan motor. As the wearer ventured into the sun, the fan would start up and blow a cooling stream of air in his face. Except for the difficulty of smoking while wearing it, it was a great idea. I wish I'd thought of that.

Ed Jackson was also personally responsible for a life-long problem of mine. It all began innocently enough in the spring of 1956, when Mary Ruth and I were invited, along with the Nygaards and others from TI, to one of Ed Jackson and Doxie's famous backyard shrimp boils. There are worse ways of spending an evening than sitting around peeling and eating spicy fresh-boiled jumbo shrimp and sucking on a beer. Trust me, it was good. And while we were enjoying such delicacies, the conversation turned to Ed Jackson's latest project—rejuvenating an almost-new MG-TF roadster that had been rolled by its too-enthusiastic first owner.

Ed had neatly hammered and rolled out the mashed fenders and bodywork and replaced the shattered windshield. Except for the alligatored paint job from the straightening, it looked pretty good. Having never seen a real genuine English sports car up close, I was intrigued by the strange-looking vehicle. Ed saw my interest and tossed me the key. "Take it around the block!" he said.

And I took it around the block and I was smitten permanently and forever with the sports car bug. There is no known cure, not that I ever looked for one.

I also became a "Product Engineer" on several production transistors, which in hindsight I will suggest was the worst job I ever had, or maybe just the most frustrating. The job consisted of being responsible for a few types of transistors from the silicon crystal to the finished and out-the-door product. The new expression of "herding cats" pretty well described it, at least the way I tried to do it. When I finally had enough punishment and was allowed to go back to designing equipment, I swore I'd never be a Product Engineer again.

My desk/workbench/hutch, was at this time located alongside the experimental "pilot line" where new transistors were developed. It was a very stimulating place to be. Red-hot furnaces filled with hydrogen were always stimulating.

Once I was idly watching from my desk as the pilot line technician loaded a graphite "boat" of transistors into the quartz tube of a small furnace as he did a dozen times a day. He refitted the end plug, turned a valve on the gas panel behind the furnace, and then turned away, closed his eyes tightly and put his fingers in his ears. I took this as a bad sign.

"BOOM!" The furnace exploded with the sharp, hard concussion of hydrogen, showering the area lightly with quartz fragments. The tech took his fingers out of his ears, opened his eyes, and began cleaning up the broken pieces as if this was an everyday occurrence. "What the hell caused *that*?" I asked, still a little shaken.

"Well, I forgot to purge the tube with nitrogen after I had it open and it was full of air. I turned on the hydrogen and realized what I had done and that the burn-off burner was going to light it off, so I just got out of the way." He added, "I need to be more careful." I agreed.

Later that month a similar occurrence brought us a famous rallying cry, "Raise down the hy dro´ gen!" Yet another furnace on the pilot line had inadvertently dismantled itself in an explosion, causing minor damage and starting a small fire. A new lady technician from Germany, somewhat lacking in the English language, jumped up on her chair and shouted to the world, "Raise down the hy dro´ gen!" with the accent on the "dro" and a hard "g" in "gen." Maybe you had to be there. But we liked the phrase a lot and used it every chance we could for years afterwards.

In the fall of 1956, Mary Ruth and I, now with two little girls (I didn't have to work *every* night), decided we wanted to move out of town, away from both sets of our parents. Out on our own, so to speak. I liked working for TI, and decided we should try for a transfer to the Houston Technical Laboratories, now a subsidiary of Texas Instruments, in guess where.

HTL had a group designing and building industrial instrumentation equipment, which sounded sufficiently interesting. One of the bosses was Ralph Dosher whom I knew from his Lemmon Avenue days. I gave him a call and laid it on the line. "I want to get out of Dodge, er, Dallas. Got room for another engineer?"

"Yep," he said. "Pack up and come on down." Transfers were easier in those days.

I started to work for Ralph at HTL on the first day of October 1956. Oh, I forgot. I had been promoted to Grade 12 Engineer back on December 1, 1954, just like I'd hoped.

Chapter Four

And so in the fall of 1956 we rolled into Houston. met the moving van at the Bissonnet Plaza Apartments on the corner of Buffalo Speedway and Bissonnet. and moved in. It was a convenient place to begin living in Houston. being only about two long blocks down Buffalo Speedway from the Houston Technical Laboratory building. Not that I had any intention of walking to work.

Our meager supply of furniture fit nicely in the small apartment. and Mary Ruth and I quickly got used to apartment life. Our good old 1954 Ford station wagon had a covered parking place in the apartment lot that backed up to the Dr. Pepper bottling plant. This parking place set the stage for an unusual problem. Having two little kids and no baby sitters triggered it. Our entertainment was mostly drive-in movies. which led to munching goodies in the car and dribbling lots of crumbs. The car developed a serious case of roaches. They were escapees from the bottling plant's racks of dirty pop bottles just across the fence. And if you've ever seen a Houston roach. you know what I mean by serious. Stripping out the floor mats and dosing the car with various deadly poisons took care of them. but I think that's the only case of car roaches I ever heard of.

It quickly became obvious that being a one-car family was a drag. just as it had been in Dallas. so I decided to fix the problem. As soon as Mary Ruth and the kids were safely on the train and headed to Dallas for a grandparents visit. I went down to Bruce Bristol. old-time motorcycle dealer in a very industrial part of downtown Houston. I was going to buy a motorcycle and I had a great excuse: I *needed* one.

After looking over the Triumphs. BSAs. and even a Vincent *Black Shadow* (which still causes a lusting in me) and finally. my bank balance. I looked further back and discovered an ex-British army 1947 Royal Enfield 350cc single. painted with yellow house paint. leaning against the wall. Now that was more like it. Since I didn't even know how to start one. much less ride it. I enlisted the aid of Bruce himself. Bruce had begun to show signs of excitement while I was peering at the pile of junk. He admitted it had "some transmission trouble" but for $80. what's a little transmission trouble? I told Bruce that if he could ride the bike around the block, I'd buy it. Bruce could ride anything around the block, and so with no serious challenge he sold me the bike and I took it home.

Rolling three hundred and fifty pounds of dead weight up the back steps and into our apartment kitchen turned out to be a bigger problem than Bruce's ride, but with the help of a neighbor I succeeded. An hour later it was being dismantled on the kitchen floor. I soon discovered an interesting feature of British bikes at this time—the clutch thrust bearing was about two hundred loose ball bearings that rained out and rolled all over the floor, mostly under the stove.

But eventually that evening, the transmission was extracted and dismantled and the extent of the trouble located. It was serious. The boss that held the main gear shaft bearing had broken off from the aluminum side casting. I would swear that nobody could possibly break off an entire boss inside a transmission, but somebody had. The gears flopped around a lot inside the case and it was a wonder that Bruce made it around the block.

So I carried the side plate and broken-off piece into the office the next day and talked to John Pachelhofer. "Mister Pac," as he was known, was an old-timer machinist and do-anything guy who had worked at Houston Technical Labs for a long time. He said, as he usually did when faced with a challenge, "Let me see what I can do." And of course he welded the piece back very nicely with his Heliarc welder.

That evening as I reassembled the gearbox I discovered that the rewelded boss was a bit tall, which was no big surprise considering it had been ripped out by its aluminum roots and stuck back on, and the case wouldn't quite go back together. A quick late-night trip to the plant and some free-hand work with the grinder and all was well. "Some transmission trouble" had been fixed. I had my own go-to-work transportation: an $80 motorcycle that ran. Won't Mary Ruth be pleased! Well, not really. But it served me well for the year that we lived in the apartments. And I finally got the yellow house paint off and repainted it a respectable black, like a British bike should be.

Lest you sneer at this single-cylinder "thumper" that I now called transportation, let me relate a tale. One of the employees had a 1956 V-8 Corvette and used to give me a hard time about my junk motorcycle. Of course this led to a match race in the parking lot behind the plant. I don't recall mentioning it to anyone, but about half the plant showed up at afternoon coffee break for the race. It was to be a short drag race, maybe two hundred feet, and I was pretty sure that his automatic transmission would be a great detriment. I was right. I beat him soundly (well, by fifteen or twenty feet) two races. Of course he was closing on me at about fifty miles an hour, but I still beat him to the finish line. He never caught on that he should have held his brakes and revved up the engine to "pump up" the transmission. And I didn't tell him.

The result of this race was that he went out and bought a 1957 Corvette, with the Duntov engine and a four-speed stick. And of course, I wouldn't race him again. He would have sucked me up his exhaust pipe with that thing. My motorcycle *was* a piece of junk.

I went to work for Ralph Dosher who was in charge of the development of recorder products. The "recti/riter" strip-chart recorder was a nice industrial product and selling well, and needed some accessories to go with it. I was assigned to design a DC amplifier. Meanwhile, a couple of events had occurred by January 1, 1957. The first and most important to me was the acceptance of my application as a Professional Engineer, State of Texas. In December 1956 I officially became #14507 with a seal and everything. The second thing was probably more important for my future well being, in that Houston Technical Laboratories, subsidiary of Texas Instruments, became Industrial Instrumentation Division of Texas Instruments. The big difference was that I again participated in TI profit sharing. Every year counted in the profit-sharing game.

The DC amplifier was finished and packaged in time for the big spring electronics show, the Institute of Radio Engineers convention at the Brooklyn Armory in New York City. It was a lot of fun. New York City was a new world and I never ceased to be amazed by its variety and vigor. This trip I helped Clark Fischel, chief TI trade show-person, set up our display and in return Clark took me out one evening to the famous Gaslight Club. Clark assured me that it was a genuine converted bawdy house. I was very impressed, having never seen such elegance in person before.

It was a new experience to ride the subways and walk the streets under the elevated trains at the Brooklyn end. Once, on a grubby street under the rattling "Els," I was singled out and stopped on the sidewalk by an elderly lady shopkeeper standing by the door of her used clothing store. After overcoming a little difficulty with the language, I got the idea that she wanted me to fix her hanging light bulb switch because she, at about five feet in high heels, couldn't reach it. I was the tallest person she'd seen walking by and so invited me into the dimly lit shop. I followed as she shuffled to the back, and at her direction I tied a new piece of string to the pull-chain on her light. As I left she nodded and thanked me. Who says New Yorkers are cold?

I arrived back at the Buffalo Speedway plant after a hard week in New York City, and was ready to get back to work. But my Houston friends had also been busy while I was away. My motorcycle, which I left in the basement of the plant for safety, had been grossly decorated with a "Welcome Home Ed Millis" sign and crepe paper threaded through the spokes and around the tires and everywhere else. I don't mean a little crepe paper decoration, tastefully done, but as done by a band of jealous engineers.

The front half of the motorcycle was almost obscured by the bands of crepe paper, and rising above it all was a flag on a stick proclaiming how glad they were to see me. Yeah, sure. Any idea of starting the motorcycle was abandoned when I noticed that the paper near the bottom of the bike was oil-soaked. What else? It was a British bike, designed and built during the decades of the Great Gasket Tragedy. The bike would have gone up like a Roman candle if a spark had come out of the engine. I was late getting home for supper.

The basement of the Buffalo Speedway plant was mostly open space, with a few special rooms for the facilities equipment, but by and large it was designed and used for automobile parking. It held a couple of dozen cars but was reduced by a few spaces every year as they thought of something else to build in the basement. The parking spaces were marked off and numbered, and carefully allotted by seniority to the few lucky employees. A list with a floor plan was published periodically by Personnel, and strict tabs kept on the basement parkers. And people like me knew just how many people had to transfer, quit or die before their turn came up for basement parking. Surprisingly, my employment time of six years was pretty high, and I kept close track of my progress. But in the meantime, since I was usually driving a motorcycle to work. I got special dispensation to park it in an unused corner too small for a car, just like the several bicyclists. Hell of a deal.

Gil Clift, a friend and fellow engineer, had done me some long-forgotten mischief, and it had been percolating in the back of my mind for some weeks that I should get even but couldn't think of anything suitable. I also suspected he had been a prime player in the motorcycle-decorating escapade but I couldn't prove it. But as I was wandering through the basement, the opportunity to get even suddenly revealed itself. In the corner I stumbled across a stepladder. Also within my view was Gil's bicycle, parked in another corner of the basement. Putting two and two together and getting a lot more than four, I used the ladder to tie Gil's bicycle to a convenient I-beam with a piece of rope. Then I *really* hid the stepladder.

I can still see Gil, late in the day with his hands on his hips and gazing thoughtfully at his bike, which was about seven feet in the air. "Hey, Gil! What's your bike doing up *there?*"

But the basement came near to being my great undoing in the spring of 1957. Never underestimate the power of a screw-up. At least *this* screw-up.

I had begun building balsa-wood "hand-launched" model gliders for a little outdoor sport during lunch and coffee breaks. The plant had lots of open space around it and chunking gliders had been one of my favorite sports at Rice. I could make one fly pretty well, and after I'd been doing this a few weeks under the envious gaze of fellow employees, others began showing up with balsa gliders. Pretty soon we had a nice group of glider fliers. So I moved up a notch in technology. I bought a "Jetex 50" solid-fuel rocket motor and attached it to a modified balsa glider. The main modification was a thin sheet of asbestos paper so the balsa wouldn't catch fire from the rocket exhaust.

The Jetex unit was an aluminum cylinder about the size of two joints of your ring finger that could be loaded with a fuel pellet and an ignitor fuse. This tiny rocket engine then fitted into a spring clip on my model and it was ready to fly. The Jetex 50 unit was the smallest they made, with a thrust of a half an ounce maybe, so it didn't exactly turn the glider on its tail and rush it off towards the moon. It just flew in a nice gentle climbing spiral until it burned out. and then would gracefully circle to the ground. Except that day.

I was standing in the lee of the building near the ramp for cars exiting the basement and getting ready to light off the Jetex for yet another fine flight. None of this crude glider-throwing for me anymore. I lit the fuse and waited a few seconds for the solid propellant to come up to full power before I launched the glider. Perfect.

I gave the spewing glider a toss and it promptly circled downward into the car ramp. The rocket wasn't up to full power after all and I had launched it too soon. Oh well.

The plane bumped the wall of the ramp as it spiraled downward and knocked the rocket unit loose from its mounting clip. Now, up to full thrust at last and freed from the weight of the glider. the Jetex unit by itself whizzed down the ramp and into the basement, leaving a corkscrew trail of smoke as it vanished from sight.

Meanwhile, Bob Olson. manager of the Houston plant. was walking through the basement inspecting the alcohol locker for fire hazards. My good old rocket did everything but go up his pants leg. This wasn't helped by my ambling down the ramp a few seconds later and saying. "Hi Mister Olson! Did you see my rocket?" Yes. he had. He certainly had.

Bob, normally the most polite and unexcitable person at TI. got pretty excited by this surprise of mine and told me so. My rocket-ship hobby came to an abrupt halt.

Later that afternoon. I headed down the hall to Bob's office to apologize for doing something really dumb. like firing a rocket into the basement. and met Bob coming up the hall. He was on his way to find me to apologize for yelling at me. What a class guy.

Just before the end of May 1957. I was called into Dr. Hal Jones' office. Dr. Jones. the Chief Engineer of the Industrial Instrumentation Division. had a problem. It seems that Dallas needed more silicon transistor manufacturing capacity and had decided to build a production line in the Houston plant. TI was going gangbusters with silicon transistors and there was some feeling of making hay (or money, actually) while the sun was shining. And it was shining plenty bright in the silicon transistor world.

Dr. Jones was aware that I was the only employee in the Houston plant with transistor engineering experience, and although my experience was with germanium transistors, it was way better than anyone else's was. He explained to me the difficult position that TI was in, not being able to free up anyone in Dallas to be product engineer. And with the exception of Jim Reese, the new manager who would be coming down to get the line up and running, everyone else would have to come from the Houston operation. So, what did I think about being the new and highly esteemed silicon transistor product engineer? I told Dr. Jones that I couldn't think of a job in the whole world that I would less like to do, and that I was perfectly happy working for Ralph Dosher on recti/riter stuff. But thanks, anyway. I was dismissed from Dr. Jones' office.

I was called back to Dr. Jones' office an hour later and notified that I would be the product engineer for the 900-series silicon transistors in Houston beginning 1 June 1957. I said. "Yes, *sir!*"

I spent a hectic three and a half months being product engineer for the silicon transistor manufacturing line. The good news was that Jim Reese was a fine man to work with and we got along great. From the disaster of the anti-vibration coupling on the oil-free air compressor to the day someone mixed up about five thousand dollars worth of Silastic-filled units with the regular production, we stayed frantically busy. Like most transistor production, the work segued from one disaster to another, with panics in between. Things went pretty normally.

Probably my biggest on-going problem was with the line foremen and line operators being creative. Some days it just seemed like they needed to try something new, and "see if it wouldn't be better to do it this way." It was hard enough to build silicon transistors if you did it just exactly like the cookbook said, much less by changing times, temperatures, helium flows and any other knob they might find. Threats had little effect, since they were convinced that creativity and finding better ways to do their job showed great initiative and should be rewarded, not punished.

One of more creative ladies on the production line had also developed a knack no other person could manage. She could go to sleep while leaning on her binocular microscope with both eyeballs. It was very hard to tell when she'd dozed off. It was with mixed feelings that I read in the morning paper that she had been arrested when she was seen throwing a cash register out of a car while being chased by the Houston police. She was being creative again but the police were even less than impressed than I was.

On the 16th of September I transferred out of Transistor Products back to where I had come from, namely working for Ralph Dosher. My ordeal as product engineer was over, and it hadn't been nearly as bad as I thought it would be. Of course I had thought it would be the absolute pits, but working for Jim Reese had made it almost bearable.

The recti/riter product line never looked better, and I gratefully jumped back into the world of engineering design. One inspiration I had for low-cost chart paper for our recti/riter customers didn't work out. After having paid what I thought was an exorbitant amount for rolls of chart paper for our engineering lab recti/riter, I mailed the TI recti/riter sales office a roll of sanitary paper toilet seat covers. The roll was just the same width as the recti/riter paper, and it only cost a buck or two versus fifteen for the chart paper. I got a memo back from Sales, thanking me for the idea, but they found that the sprocket holes were way too big. Darn.

But in the spring of 1958 I had a success. On April 1, I sneaked out of the building and installed a crudely painted "black box," complete with tripod and "cable" dangling down, alongside Buffalo Speedway. This may be hard to believe today, but that was a very rough replica of the police speed-checking radar systems of the time. Hand-held it wasn't. And how anyone thought it could work with no one anywhere near it is beyond me, but the mystique of the dreaded new-fangled Police Radar was such that it slowed the traffic on Buffalo Speedway to a crawl all afternoon. There was even a small piece in the Houston paper the next morning:

> *April fool pranksters kept traffic on Buffalo Speedway between Richmond and Bissonnet moving at a snail's pace for almost five hours Tuesday. A small black box atop a silver tripod that looked like a radar unit police use to clock speeders had been placed on the grass near Buffalo Speedway between Sohio Petroleum Co. Bldg., 3813 Buffalo*

*Speedway and the Texas Instruments, Inc. Bldg, 3609 Buffalo
Speedway. Motorists slowed down, honked at drivers behind
them to give warning. Police said no radar unit worked that
area Tuesday.*

Believe it or not. TI got a call from the Spring Branch Police
Department later in the week asking if we could build them a couple
of fake radars, since they seemed to work as well as the real ones. TI
declined to add it to their product line.

It was about this same time in 1958 that I got a call from my
old friend and boss, Jim Nygaard in the Dallas transistor test
equipment group. He'd been thinking again. Would I like to run a
group in Houston reporting to the Transistor Products in Dallas?
They needed a lot of stuff designed and built in a big hurry, and a
free-wheeling group. far enough away from the maelstrom of the
Dallas plant just might be a good idea. I thought about it for a
microsecond or two and said hell. *yes!* I liked working for Nygaard,
and the word "free-wheeling" had a nice sound to it. It was time for
a change.

The details of how this Houston arm of the Transistor
Products test equipment and mechanization group was set up are not
clear to me. All I remember is that Nygaard twisted enough arms to
cause Houston to do a nice double back-flip. Houston was going to
set up a group. with me as the lead engineer. reporting to Nygaard in
Dallas through Ralph Dosher on a dotted-line basis. And every other
group in Houston had to contribute personnel to staff it. Hot damn!
What a chance for everybody to get rid of their troublemakers!

I'm sorry that I don't have a copy of the roster of that first
group. Nearly everyone I got had been a management pain in the ass:
misfits. troublemakers and rule-breakers. It was exactly the kind of
group that Nygaard wanted to put together and it was the best team I
ever had the pleasure of working with. Not that they didn't drive me
crazy periodically. but that was a minor annoyance compared to their
talent and energy.

The process by which this team was put together, namely
managers taking the opportunity to unload their unsatisfactory
employees on a start-up group. worked because of one reason—they
were unsatisfactory employees not because they were dumb. they
were unsatisfactory because they were smart. They weren't satisfied

with the way things were and wanted to change things, and because they had better ideas and nobody listened, and because they didn't fit into the eight-to-five mold and because they were basically irreverent to authority. And every one of these "problems" turned out to be an asset for the Transistor Products design group. It was truly a special group of guys.

My fading memory brings up George Blackstone, Joe Rich, Ray Nixon, Gil Manire and Harold Armand. I wish I had the complete roster.

The first product our embryonic design and fab team built was an idea that I had been thinking about since being a reluctant product engineer. Some days when you were building transistors, you weren't worried about whether one was a tiny bit better than another, but what the hell kind of transistors were they, and who left this unmarked sack on the table anyway? Like were they NPNs or PNPs, and were they silicon or germanium? And were you sure the bar was installed properly and the correct leads were the collectors and emitters? Like so often happens, it's not the details but the fundamentals that get you.

I cooked up a little box with a custom-faced meter on the front and a transistor socket on top. If you plugged in the mysterious transistor that you found on the floor, this box would tell you whether it was a PNP or an NPN, silicon or germanium, and which lead was the collector. It was not a very complicated circuit; it's just that no one ever thought to build one before. It was a smash hit in Dallas. We sold a lot of them to people who couldn't keep their transistors straight, which was practically every line foreman in TI.

Our group was loosely structured and with Nygaard's broad charter, we began designing and build a few free-lance things for the Dallas Transistor group. But there was a strong vibration coming out of our past: the CAT Machine. Proposed in 1955 and rejected by management, Jim and I thought the time had come. Besides, Dallas management couldn't tell what we were doing in Houston anyway. It was a perfect set up—money, time, good people and corporate invisibility. And the most important of all, we had Jim Nygaard for a boss.

We had something else going for us that I didn't understand until much later. We were going to bootleg a large and important project, and it behooved us to do it rapidly and successfully. There's nothing much worse than being caught in a bootleg project that doesn't work. Consequently, after we looked over our first CAT Machine proposal, we began to simplify it to make it easier to design and build. One of the first ideas was to scrap the idea of a moving paper tape synchronized with the transistor carrier blocks, and just put a piece of paper on each block and let the block carry it and index it. This eliminated a lot of machinery that would have been required to move the paper tape accurately.

Jim Nygaard and I understood the complexity and uncertainty of testing transistors. Any test box designed for a particular transistor could be obsolete in a few months. Test parameters would change, and this necessitated ripping the guts out of the test box and rewiring it. We decided the CAT Machine needed really cheap, flexible, plug-in test units. Most of the test station will stay the same, but a small portion of it would determine the test circuitry and parameters. And did I say make it cheap?

The other major problem with production line transistor testing was keeping the damn things calibrated and rejecting the rejects and accepting the good ones. We agreed that the CAT Machine would use the "dead cycle," while the blocks were being moved to their next station, to check the calibration of the test sets. Each set would have a resistor placed across two of the transistor socket terminals to simulate a "just barely fail" condition. If the test station didn't register "fail" as it should, the CAT Machine was shut down and the offending test box identified by a light. Each test set checked its calibration between each transistor it tested. It was another really good idea.

So we built up a breadboard model, which was electrically and mechanically functional, but really ugly, of the main parts of the CAT Machine. We also built a few dozen carrier blocks, each of which could carry one transistor in a socket that was wired to three copper saddle rivet contacts in the top. Under the socket was a spring clip to hold a business card-sized paper card. An air-powered block "pusher" was built as well as a couple of yards of track for the blocks to run in.

Three test stations were built, each of which had a set of electrical contacts that would drag on the saddle rivets to hook up the transistors. The output of the test station went to a hole punch.

This punch, strategically placed by the track, would punch a hole in a paper card carried along with the transistor if the transistor failed the test. If the transistor was good, it didn't punch. Later we learned to tell how good or bad a group of transistors was by how much racket the punches were making.

After the three test stations were two sorting stations. These stations compared the holes in the cards with a patterned plate and decided if the transistor had passed the right tests to be selected. A pneumatic arm would jerk the transistor out of the socket and drop it in a bucket if it passed the proper tests.

Our first test of "card reading" was a failure. I had a local printing company cut up a few thousand cards out of two-ply black index stock for our trials. What would be better than black card stock to block the light? Well, lots of things as it turned out. The photosensors used in our sort stations were sensitive to red and infrared light, and the infrared light went through the black card stock like it was transparent. It was truly amazing. The signals were just about the same from a punched hole and a non-punched card surface. Panic was averted when samples of other colors were quickly tried and "two-ply blue index" cards were found to work perfectly.

The photo I have of this breadboard model looks like some sort of tabletop disaster. Pliers and cutters and pieces of wire are lying on the table and the whole area overlaid with a general dusting of chad from the punches. It resembled a pile of electronic and mechanical junk, but we thought it was the most beautiful thing in the world. It was a brand-new idea for testing transistors and we had thought of it and we built it and we *knew* the sonofabitch was going to work.

Jim Nygaard thought it looked good, too. So good, in fact, that he decided it was time to blow our cover. He told Mark Shepherd, boss of the Semiconductor-Components Division, that he had something he wanted to show him at the Houston plant. Jim was a good friend of Mark's and Mark was pretty sure that if Jim wanted him to go see something in Houston, it would be worth his time to go.

Jim briefed Mark on the way down on the plane while we prepared to give them a gee-whiz demonstration when they arrived.

We did, and Mark liked what he saw. He told Jim, "Do it!" And then he said the famous words to Jim that we quoted so often afterwards, "Why didn't you tell me? I could have helped!"

We kicked around our next step and decided that a CAT with eight test stations and four sort stations would be a useful system and would serve to prove out our ideas in a production version. And it had to look pretty. We chose "Steel Blue Wrinkle" as our elegant finish, stolen directly from the finish used on Tektronix oscilloscopes. Before CATs became obsolete, we bought a lot of Steel Blue Wrinkle paint.

My Houston group and I scrambled to get schematics and designs ready to build, and within a few days we were ready to start ordering parts and making chips in our little machine shop. But I had a problem. In perfect timing with this fire drill to build the very first CAT Machine in the world, I had to leave on vacation.

Mary Ruth and I had arranged to meet with my brother Craig and his wife Ruth for a two-week vacation. Plans and reservations had been made and I didn't want to screw it up for all of us. I also wanted to stay and build the CAT, but I went on vacation. My guys called me a lot of names as I left.

When I dragged back into the plant on the Monday morning after our vacation, every person in my group was sitting at his desk or workbench, with his feet propped up and reading a magazine. To a man. Then I looked in the adjoining area, and there, on three tables bolted end-to-end, was the completed CAT Machine, Steel Blue Wrinkle paint and all. It was gorgeous! Those crazy sons of bitches had worked night and day for two weeks just to show me that they didn't need me to build no stinking CAT Machine. I was overwhelmed. I thought it would take at least a month to get it together because of the machine shop work and special parts. But they found a way. And then they couldn't wait until I came in Monday morning. It was one of the finer surprises of my young life.

The prototype CAT was moved to the Dallas Semiconductor plant for production testing and it ran with no major problems. It tested 1200 transistors per hour, with each transistor receiving eight different electrical tests. Three people could run this small CAT. One person loaded the transistors into the test blocks as they passed by while a second person inserted the paper card in the block clip.

A third person would walk the line and empty the sort buckets and carry the test blocks back to the beginning of the line.

The CAT replaced ten or a dozen test persons and did a more reliable job. The only negative comment that I recall was a Dallas manager that said it wasn't going to work because there was no good way to get the test blocks back to the beginning after they had passed through the machine. At the time we were using the bushel basket transport method which I suppose looked a little crude. But fortunately, someone in our group remembered that the conveyor belt had been invented and we were soon off the hook.

On June 27, 1958, Harold Armand, ace electrical engineer from my group, Dick Arnett, department manager in the Houston plant, and I met with various product groups of the Semiconductor-Components Division of Texas Instruments in Dallas. We met for the purpose of allowing Mark Shepherd to browbeat his managers into buying CAT machines. As a July 2 memo from Dick Arnett said:

> The meeting was called by Shepherd so none of the product managers could ignore the situation but they were placed in a rather difficult position of being asked to make a decision on very short notice and with very little information. Inasmuch as Shepherd was enthusiastic, various groups did agree to place some specific requirements with us at this time...

Dick phrased it very politely, since it was a railroad job from start to finish. And Mark's bullying was proper—he needed to get CATs into his production areas as rapidly as possible. The final tally from the less-than-enthusiastic future CAT owners was four complete machines with a total of 60 test stations and 30 sorts from the IBM Transistor Products bunch, and one CAT with 20 test and 5 sort stations from a germanium group. It was estimated by Dick Arnett that this would be about $80,000 worth of equipment, although this later proved to be low by a factor of two. The Houston management was delighted by this sudden surge of business from our group. They even quit complaining about the space we were taking up.

Even my old nemesis Gil Clift had something to add to the CAT program. He suggested changing the name from "Centralized Automatic Tester" to "Pushing Under Semiconductor Station Yanker," or PUSSY. We were not amused.

But for all the work, I was having fun, too. Back in March of 1958, about the time we were setting up our new test equipment group, Mary Ruth and I were invited over to Bill and Rosy Illingworth's house for dinner. We goo-gooed at David, their new baby, ate good food and had a fine time. As we were leaving, I noticed a strange-looking dilapidated green convertible automobile of some sort in the backyard next door. Ever drawn to broken-down machinery, I got the name of the neighbor from Bill and called him the next day. I arranged to go by and take a closer look at this "Singer," as he called it, and to see if it would be an interesting restoration project. Not that I'd ever restored a car before, but what the heck?

We had moved out of the Bissonnet Plaza apartments and into a house in the Westbury section of Houston. I had sold the motorcycle, as it was way too far to ride to work, and we were struggling as a one-vehicle family again. To my optimistic eye, this pile of rubble looked like a possible replacement. And believe it or not, it looked a little like the MG-TF that had so lethally infected me with the sporty-car bug in the spring of 1956. Just a little, but enough.

The car had been sitting in the back yard for several years and looked it. The convertible top, such as it was, had been folded down and a ratty tonneau covered most of the seats. The left rear fender, ripped from the body and considerably wadded up, had been tossed in the front seat. It was a 1952 Singer SM1500 Tourer, which meant a four-seater convertible. It was British-built, and it was in just the right state of disrepair, like total. For $125 it was mine.

Months later, I had cleaned it up enough to run and undertook to drive it around the block on its maiden voyage. Well, not exactly its maiden voyage, but its first voyage in several years. I cranked it up, backed out of our driveway and headed out. It smelled a little smoky, but I attributed that to its lack of recent use. After a circuit of our block I pulled back into our driveway and glanced down the street from whence I'd come. It looked like the Houston Mosquito Patrol had just finished a major fogging. A pall of white smoke hung over the whole neighborhood. Houston, I had a problem.

In another month I had extracted all four pistons from the car and observed that the oil rings had been installed wrong. For you auto mechanics, let me add that getting a piston out of a Singer was not as simple as it might sound. After the usual head, pan, and rod bearing cap removal, the piston and rod were shoved up from the bottom for removal. Except the big end of the rod wouldn't go through the bore. So you held it there with one hand, while with your other two hands you supported the piston and tapped out the gudgeon pin. Simple as that. But back together again with the oil ring bits in the right order it ran fine, with no more smoke than the usual British long-stroke, oil-oozing engine.

One bit of Singer Serendipity occurred while I was browsing a junk/antique shop on Westheimer one fine Saturday. Outside, in the serious junk section, I peered into an old wooden nail keg, and did a double-take. I reached in and pulled out a Solex 30FAI carburetor. Guess who needed an intact Solex 30FAI carburetor top casting to replace a broken one that had been epoxied? Me, that's who. As I stood there, stunned by my good fortune, up walks the shop's owner. "How much?" I asked, trying to keep a straight face while extending the carburetor with my left hand, my right hand casually in my pant's pocket.

"I hear some change in your pocket. Give it all to me and whatever that is, is yours." The carburetor cost me 87¢.

Roger Webster from my ASQ-8 days was kind enough to drop by a wrecking yard in California, the source of all strange car parts, while on vacation that summer and bring me back a used Singer left rear fender. A professional paint job of British Racing Green and a new convertible top followed the installation of the fender. We were in business. Except for the usual Lucas Electric problems, tightening the wood screws that held the door hinges to the ash frame struts occasionally and checking for termites, it was an altogether satisfactory car. After a hard day at the office, how many other cars can you lay the windshield down flat on the hood and get the wind in your face on the way home? Not many.

And blessedly, by this time I had achieved the enviable perquisite of a basement parking place. Since the Singer was not exactly hermetically sealed, it was good to keep it out of the Houston weather.

Then one day after work I walked down to my basement parking spot and the Singer was gone. Not just moved, but gone. I searched the basement and then the outside lots to no avail. I couldn't imagine anyone wanting to steal a Singer. Not without the shop manual, anyhow. I had gone back into the basement and was standing around looking worried and trying to decide on a course of action when a sympathetic friend casually winked at me as he passed and said, "Was that pile of boxes there this morning?"

Cleverly concealed in the corner of the basement under and behind probably a hundred large empty cardboard boxes was my precious Singer. Somehow I never thought to look there.

The deed was undoubtedly done by the same person or persons who jacked up the back axle a few weeks later. A cinder block was slipped under the axle to raise the right rear tire just clear of the ground and prevent the car from driving off. I couldn't imagine why it wouldn't go—engine running, in gear and the clutch out—but not moving an inch. I hate to admit how long it took me to figure that one out.

Parallel to our efforts of getting the CATs into the Houston production schedule, Dallas brought up another testing sticky wicket. How could we make the CAT do testing at elevated temperatures? IBM was particularly tough in their specifications for computer transistors and required 100% testing at 55 degrees Celsius. Harry Waugh, in Nygaard's Dallas group, was given the challenge. Remarkably soon, Harry had a "wind-tunnel oven" installed and running like a champ on the prototype CAT. The test blocks would progress down the CAT track for the room temperature tests, and then enter the heated tunnel. After a short section of tunnel to bring the transistors up to temperature, they would reach the 55 C test stations. The final versions of the temperature-test CATs were forty feet long and worked to perfection. This was the blessing of modular construction—we had no idea what a CAT would need in the future, but we left room.

As another sidelight to this development program on the CATs, in September 1958, Dr. Warren Rice, Professor and Department Head of Mechanical Engineering at Arizona State University, was put under contract to TI and Jim Nygaard. He was to develop a compact source of cold air that we could adapt to the CAT for low temperature transistor testing. After some study, he chose the simple Tesla bladeless turbine to use as a method of expanding compressed air to lower its temperature. His test models produced lots of cold air from lots of compressed air, but the requirement for low-temperature testing mercifully faded away and it was never needed. However, Dr. Rice developed a life-long interest in the Tesla turbine and was the first to analyze it mathematically. He's mentioned in the book on Nikola Tesla's life, "Tesla, Man out of Time" by Margaret Cheney, as "an authority on Tesla's work in turbines and fluid mechanics..."

And of course, little did we know what Jack St. Clair Kilby, in an obscure lab in the new Semiconductor building in Dallas, did on September 12, 1958. He hooked up his tiny experimental phase-shift oscillator to a ten-volt power supply. The oscillator was built on a single sliver of germanium and it oscillated. This was the oscillation that was to be heard around the world. Jack had invented the integrated circuit.

Page 20 of Jack's engineering notebook of that day begins, "A wafer of germanium has been prepared as shown to form a phase shift oscillator." The entry ends, "When 10 volts were applied...the unit oscillated at about 1.3 Mc, amplitude about 0.2 v pp..." This was truly a world-changing event. Thanks, Jack.

This, of course, brings to mind a story I heard about Jack a few years ago. In October of 1983, Jack was in his office and on camera and being interviewed by Diane Sawyer. As she was winding up the piece, she pointed to the convoluted hen-scratching and diagrams on the blackboard and asked Mr. Kilby if that was his next invention after the integrated circuit. No, said Jack. It was the directions to the garage that was repairing his car.

The CATs were off and running, so to speak. A big open bay in the new Semiconductor Building on the Expressway Site was soon filled with them. I can still hear the rhythmic din of those machines, each running at its own frantic pace. It was tremendous. The pulsing and throbbing seemed almost alive as it swelled and faded, on and on, pounding and pulsating. Her scarlet lips parted as she turned towards him... Wait a minute. That's another story.

71

Early in the rush to build CATs, Textool Products, a machine shop in Irving, Texas, was contacted. Sam Braun, a machinist who had worked at TI in the early transistor days, had started this up-and-coming shop. To say he was a machinist is to call Beethoven a music writer. His work was elegant and flawless and he expected everyone who worked at Textool to turn out the same quality of work. He was from Germany and had been in the German air force in the First World War. He was a mechanic in an observation squadron and I can still hear him telling about being required to go for a ride in the plane he had just worked on. He would say, "You talk about your qvality control!" in his German accent.

And to top it all off, Steve Karnavas, the TI transistor production boss of a few years back, was now the manager of Textool. Textool had an unblemished quality and delivery record, and we personally knew the top guys. Dealing with Sam and Steve was easy—give them a job and they did it properly, on time and charged a fair price. We didn't have time to get three quotes and haggle; we just wanted to get CAT machines built properly and in a hurry. We picked the right place because they did just that. They built a ton of CATs and CAT parts over a three or four-year period and we were never disappointed.

The Houston plant was working hard too, building our designs for the Dallas transistor production areas. Buford Baker, a mechanical engineer, was the manager of the Houston production department. Buford ran the facility with aplomb and grace. He was never excited and always ready to listen, and the employees liked him a lot. Even to the extent that Buford told me one day that he wished they didn't like him quite so much. One of the ladies on the assembly line had come into Buford's office in a Grade-Three Snit one afternoon and confided to Buford that she was going to kill old What's Her Name on the line if she didn't quit messing with her husband. And then she opened her purse and showed Buford the pistol she was going to do it with.

Buford allowed as how they needed to go to a higher authority with that problem and asked her to first close her purse and then follow him up to the office area. Buford, who didn't get along with the then personnel director, escorted his gun-toting employee up to the personnel director's office, pushed her in with a "This lady needs to talk to you!" and closed the door and left.

Another of Buford's line ladies got in a bit of trouble one day in a manner reminiscent of the beans-in-the-burn-in-ovens problem in Dallas. As I was passing through the assembly area at lunchtime, I heard a muffled pop and a scream. I headed over to see what the problem was, but the area was already deep in a cluster of solicitous women. I was waved off and told in no uncertain terms that I was neither needed nor wanted. The problem, I found out later, was that a lady was heating a can of chili with a soldering gun. Her mistake was in gripping the can between her knees underneath her skirt and holding the hot soldering gun to the lid. It exploded, blowing the lid off and directing the chili in a manner I don't want to discuss. I'm glad they stopped me when they did.

At some time during the era of the CAT machines, Chester W. Nimitz Jr. had taken over as the manager of the Houston plant. He was, as you might have guessed, the son of Admiral Chester W. Nimitz, Commander of the Pacific Fleet during World War II. Chester Jr. had been a submarine commander and was one salty individual. He and Jim Nygaard hit it off immediately, as both were chomping at the bit to get things done and their patience with big-company red tape was negligible. Chester was a big help to Jim in his dealings with Houston. He stood without ceremony and full speed ahead was his standard operating mode.

One day Jim Nygaard, in Houston on one of his regular visits, and I were leaving the plant for lunch when we met Chester coming up out of the basement, two steps at a time, with his lunch. His lunch consisted of a cellophane-wrapped ham sandwich from a vending machine. He and Jim immediately engaged in a vigorous business conversation while I stood by watching. Chester was eating his thirty-five cent sandwich during this exchange, and in a particularly energetic exclamation a large piece of ham flew out of his mouth and landed on the lapel of Jim's suit coat. I was horrified. I tried to imagine what Emily Post would recommend in a case like this but I couldn't. It was a hopeless faux pas. Wrong. Chester, never missing a word of his discussion, reached over, picked the piece of ham from Jim's coat and ate it. And Jim didn't bat an eye either. In fact, I don't think either of them even noticed the incident.

Chester was known to barrel down the narrow front hall of the Buffalo Speedway plant yelling "Gangway!" There was also a story of his accidentally barging into the ladies' restroom and asking his surprised secretary, who was in the midst of pulling up her drawers, "——? What the hell are you doing in here?" This convinced her that she had erred and not Chester as she ran red-faced and flustered out into the hall.

But when I tried to stir up Ralph Dosher's secretary with a little spur-of-the-moment joke, it didn't work out at all. For some reason I had dropped by her secretarial location while carrying a Christmas tree light flasher. Not finding her at her station, I suddenly realized that I could brighten up her otherwise dull day by connecting her electric typewriter to the flasher. So I unplugged her typewriter and installed the flasher unit in the socket.

Of course, I had to hide around the corner and watch when she finally came back to her office, turned on her typewriter and began to type a memo. About ten seconds into the memo the typewriter motor quit and the typing stopped. Puzzled, she paused long enough for the flasher to cycle again, and the typewriter started up, and so did she. After a minute or two, she got the hang of the starting and stopping, and had timed her typing to fit the spastic typewriter and was making great progress on her memo. She thought it was kind of fun. Big joke failure.

Sometime in 1958 someone mentioned to me they'd heard a person in Bell Labs had built a box with a label saying, *Do not turn this switch on!* When the switch was turned on, the top of the box opened and an arm would come out and turn the switch back off. That was absolutely the funniest thing I'd ever heard. My God, I wish I'd thought of that! But even if I didn't think of it I could recognize a terrific idea when I heard it, so I built one.

We had a lot of fun putting it around the plant and goading people into turning it on and then watching the different reactions. The totally unexpected arm emerging from the box really freaked people out. They were either frightened or laughed themselves silly.

I wrote it up and sent it in to *True* magazine and got a check for $200 when they published it, which put it far and away at the top of my list of money-making jokes. The box, which by now had acquired the name "Power Saver," was featured in a couple of articles in technical magazines, and a piece in the TI newsletter. I got several letters forwarded to me from people asking for plans.

The penultimate compliments were two requests for a loan of the box for display in modern art museums, which it weathered successfully. The ultimate compliment was when the Power Saver was stolen out of the Buffalo Speedway plant lobby a year or two later. I still think it's a funny idea and wonder who it was at Bell Labs that originally thought of it. My hat's off to you, sir or madam. You truly have a warped sense of humor.

Speaking of moving arms, it was also about this time that we decided to build a motorized Finger Machine for Boyd Cornelison. This machine could automatically give someone the finger merely by stepping on a convenient foot switch. We didn't want Boyd to sprain his hand or anything since he gave a lot of fingers in the course of a week. The "finger," with attached hand, was simple. Some rude company made a ceramic ashtray out of a full-sized hand with properly extended middle finger. The ashtray was then attached to a motor-driven arm that propelled it from inside a box in the proper upward arc. In fact, we built two of them, since Mark Shepherd had been known to use the same gesture when displeased. But we modified Boyd's to be the deluxe model—when the foot switch was held down, the finger would oscillate up and down just like Boyd was doing it personally. Boyd loved it and kept it on his desk for years. It was a nice piece of work if I do say so myself.

Mark sprung his new toy on the company directors at a board meeting. The mysterious box sat on the table with no explanation for most of the meeting until someone said something Mark didn't like. The story goes he calmly turned the box, aiming it at the offending person, and stepped on the foot switch. Mark's secretary, outside the door at her typewriter, told me later that the roar of approval was tremendous.

Strangely enough, the paranoid secrecy that the Semiconductor-Components Division maintained about the CAT machines was dropped and I was allowed to present a technical paper at the Institute of Radio Engineers 1959 Spring Convention and Conference in New York City. As I sat on the front row in the auditorium waiting for my presentation time, a box of slides clutched in one sweaty hand and my notes in the other, I struck up a conversation with Claude Head, sitting on my right with his box of slides and notes clutched in *his* sweaty hands. Claude was from Varo Manufacturing Company in Garland, Texas and also presenting his first paper in public, and we felt a certain kinship from the mutual fear. I mention this because Claude later hired into TI to a long and fruitful career and a long-lasting friendship with me. We both gave our papers without actually dying from fright, and the CAT paper was published in the 1959 IRE National Convention Record for all to see.

I note from this paper that the production speed of the CATs had been increased by this time from 1,200 units per hours push rate to "Usual test quantities are 10,000 to 12,000 units per [8-hour] shift." This translates to a push rate of 1,600 or 1,700 per hour. Later in the paper I noted a "best performance" in throughput that was occasioned by a rivalry between the day and night shift personnel resulted in a total of 16,256 transistors being tested and classified in one shift. This meant the CAT ran at an *average* speed of about 1900 units per hour. I also estimated that it would require 55 operators using hand equipment to test this volume of units. And, just for the record, I stated in public:

Test condition accuracies of 0.5% are easy to achieve and maintain over long periods of time.

Probably relating to the decision to allow the IRE paper to be given and published, TI filed a patent application for the CAT on September 10, 1959. It was a strange application in that it described and pictured the very latest in CAT technology, including Harry Waugh's wind-tunnel oven, but the inventors were only those on the original CAT proposal from 1955. The proposal was the same in principal but quite different in detail. The original inventors as listed on U.S. Patent 3,039,604 are Richard (Dick) L. Bickel, Wendell G. Brooke, Winthrop (Win) J. Day, Earl D. (Mac) McDonald, Jr., Edwin (That's me, Ed) G. Millis and James (Jim) L. Nygaard. It was issued June 19, 1962, with 11 claims.

When vacation time arrived in the summer of 1959, I found my group had not forgotten that I ran out on the previous year's great CAT fire drill. They were lurking in the lobby as I headed out

on my last day of work and they handcuffed me to the steel
grillwork. I didn't know they had any handcuffs.

     I spent a lot of time shuffling back and forth between the TI
plants in Houston and Dallas. Braniff got our business if I didn't
drive and take the family or deliver equipment. One Braniff flight in
particular that fellow engineer Everett Hanlon and I took remains
fresh in my memory. It all began on a foggy morning at the Houston
Hobby airport. We soon found out that our flight had been delayed
because the field was socked in and was not scheduled to leave for a
couple of hours. In place of arriving at the Dallas plant in the middle
of the morning, we'd be lucky to get there in time for lunch.
     We peered out of the terminal window at our non-flying
flying machine—it was a Lockheed *Electra*. Oh, swell. The *Electra*s
were setting records for coming apart in the air and killing everyone
on board. One had crashed on the Houston-Dallas run not many
months before. So we found some comfortable seats and coffee in
the waiting room of Houston Hobby and waited for our flight to be
called.
     Finally, *finally*, our Braniff flight to Dallas was called and
we hustled down to board. Glory be! It had been magically changed
into a good old reliable Boeing DC-7! This must be our lucky day! A
truer statement had never been made.
     We got aboard and found a couple of seats and settled in. The
four engines cranked up and we taxied out and began a routine
takeoff, headed for Dallas at last. The plane roared down the
runway, and the DC-7 did indeed roar, and began to lift off as it
reached flying speed. Just as it broke ground, all four engines shut
down. Two seconds later, with no noise but the sound of the wind
rushing by the fuselage, the nose of the DC-7 was jerked up
violently, followed immediately by the accelerating thunder of the
engines as they were suddenly jammed to full-throttle. The plane
shuddered and then held, and slowly began to climb out of the
airport. It made a wide circle, lined up on the runway we had just
left, and landed again.
     As we taxied back to the gate, considerably shaken, the
stewardess came on the intercom and said there appeared to be a
problem with the nose wheel and the pilot had decided he wanted to
check it out. Yeah, right. Trouble with the nose wheel. We parked at
the gate we just departed from, filed out of the plane, down the

movable stairs and onto the tarmac. As we straggled back into the terminal, one of the passengers said excitedly, "Look at the propeller on that engine!" The normally rounded-tip four-bladed prop on the number two engine was now square-tipped. It was shorter than the other three.

A few phone calls to our ex-next-door neighbor in Farmers Branch, a Braniff flight engineer, and the story unfolded. The pilot and copilot on our almost ill-fated flight had never flown together before. During the takeoff run, the pilot was watching some engine function and didn't like what he saw, and just as the plane broke ground, he decided to abort the flight and pulled back the throttles on all four engines. Meanwhile, the copilot, without permission from the pilot, had decided that the plane was up and flying and had flipped the "Gear Up" switch to retract the landing gear. And then the pilot chopped the throttles.

The plane began to settle back on the runway when the pilot suddenly realized there were no wheels under it, possibly brought about by the sound of the prop hitting the runway. He jerked the yoke back to keep from augering in and at the same time fire-walled the throttles in a desperate attempt to keep the barely-flying plane in the air. It worked. Whatever he did was just right. It really *was* our lucky day. Everett and I hadn't realized just how lucky it was going to be.

Braniff later foisted off two of the oldest and tiredest Lockheed *Constellation*s in the world on the Dallas-Houston route. This matched pair of Connies became known to the regular fliers as *Fear* and *Trembling*, although I couldn't tell them apart. The upholstery was tattered and the paint rubbed off every touchable surface. They were worn out.

One memorable flight on either *Fear* or *Trembling* was a return flight to Houston late one afternoon. As I walked across the tarmac to climb the stairs into the plane, I noticed the pilot doing his preflight check. He was standing under an engine, hands on his hips, looking up at the pencil-sized stream of oil pouring out and forming an ever-growing puddle beneath it. Friends, I'm not talking about a little oil drip, this was a solid stream. I got on board anyway.

But surprisingly, the door was soon closed and the engines started. We taxied out, took off and began the climb to altitude. After a few minutes of climbing we leveled off in the usual fashion, which was followed immediately by the shutting down and feathering of the leaking engine. This was not the biggest surprise in the world to me. The pilot informed us on the intercom that he'd discovered that one of the engines was using a little oil and he decided to save it for landing in Houston. Not to worry.

And so, as we approached Houston Hobby airport, the dead engine was lit off again, and we landed uneventfully with all four props spinning. As I understand the FAA regulations, if you land with N minus one engines running, all the fire wagons come out to meet you and it costs your airline money. However, if you can get that sucker running again before you hit the landing pattern, you're home free, so to speak. It was a great day when they finally scrapped *F&T* and got real airplanes for the route.

Interspersed with the flights back and forth between Dallas and Houston, were the dreadful drives to deliver equipment. Not trusting ordinary shipping companies and reluctant to spend the extra few hours they would require, we trucked our new and improved goodies to the Dallas plant ourselves. By ourselves, I really meant by myself. I made the tactical error of taking the bait and getting a TI-paid-for commercial driver's license. I was now Engineering Branch Manager and DTD (Designated Truck Driver). And the worst was that the "truck," using the term loosely, was Houston Technical Laboratory's old worn-out delivery van. This so-called vehicle was of the type that I associate with milk delivery in the suburbs. A seat on a pedestal, sliding doors in the front, and room in the back for either milk cartons or sophisticated electronic equipment. But no air conditioning.

As I was leaving Dallas in this *wunderwagen* late one sweltering summer afternoon on my way back to Houston, I rolled to a stop at a red light in the southern part of Dallas. Long before Highway 75 swooped through Dallas and merged into 145, the route to Houston was an endless series of surface streets through the rattiest part of industrial Dallas. And here it was that I sat at the red light, with no air moving and slowly frying in my mobile bake oven. Just to my left was a city worker, incredibly on his hands and knees in the blistering heat on the asphalt, carefully painting stripes for the left-turn lane. Since I had my sliding door wired open in an attempt to stave off heat stroke, I had a front row seat to this activity. I spoke to him, "Say, you're doing a nice job on those lines."

He looked up, startled, and I thought he was going to bust into tears, "Really? *Really?* I try so hard and everybody just yells at me for being in their way. Thanks!" I had done my good deed for the day and didn't even know I was doing it.

The year of 1959 and the first half of 1960 brought frantic activity to our Houston group. The CAT machines were an ongoing project as we upgraded the performance and added new capabilities. We branched out into whatever Jim Nygaard found the transistor production people wanted or needed. We built furnaces, fixtures, special test equipment and a variety of other gadgets.

Between moments of hilarity, we worked our butts off. Some of us worked every evening during the week, always on Saturday and frequently on Sunday. Dragging in on Sunday morning was the worst. We had a Sunday routine of borrowing one of TI's big coffee urns and carrying it across the street to the Toddle House and putting a couple of bucks worth of coffee in it. That helped a lot.

But the Sunday morning Jim Nygaard was due in from Dallas, the Toddle House was closed and we had no coffee in the urn. Jim was driving down and due momentarily and his first living act on his Sunday trips was to go straight to the urn and pull a cup before he even spoke to us workers. George Blackstone had an idea, "I've got a couple of warm beers in the car. I could pour them in the urn." Good idea, we all agreed, and George proceeded to do as he suggested.

Jim arrived and right on schedule picked up a cup and went to the urn. As it filled, he said, "This looks like beer." He took a sip. "It is beer." He then finished filling his cup with warm beer and without further ado we proceeded with our day's business. Not quite the reaction we were expecting, but it kept him from complaining about the lack of coffee.

One diversion from the transistor production equipment was the DIAL System. This acronym meant Digital Information Automatically Logged, and it was a pretty advanced system for 1960. In addition to hoped-for use in the Semiconductor-Components Division, it was a possible product for the Houston group. The problem it was designed to solve was the complex and error-prone method of keeping track of material transfers in a manufacturing operation. At that time, material transactions were written on a paper form of some kind and the data then later manually keypunched into computer cards. The cards were sent to a computer for data manipulation. The DIAL system used a multitude of remote terminals at the transaction points connected to a central IBM Printing Summary Punch, which eliminated the paperwork and the time lag. But even with the help of an Industrial Designer on the DIAL System Station, which looked really fine, the system never sold and was never used except experimentally. Not everything we did was a CAT. Some were dogs.

By May of 1960 I was beginning to tire of the Singer, what with the responsibility for termite control and such, and had begun looking around for other and more reliable wheels. I liked the convertible idea from driving the Singer, and with the generally mild (if sometimes damp) Houston weather, I had a high yield of top-down driving. So I was pleased when I saw an ad from a car dealer advertising a 1958 Morris Minor 1000 convertible for $850. Like a lot of used car advertisements, it sounded too good to be true.

The car lot was just a few blocks from the TI plant so I dropped by on my lunch hour. It was a nice-looking Morris with no damage and an intact top, which puzzled me even more with the low price. The salesman, with double-breasted suit and slicked-back hair, assured me it was the best deal in the universe. "I'd like to take it for a spin. Okay?" I asked, suspecting there was more than met the eye with this piece of British iron. He reluctantly agreed and handed me the keys.

The expected hidden problem was immediately apparent. The car didn't have enough power to pull out of its own smoke, as the saying goes, even though it wasn't smoking. I struggled to circle the block and not get run over, but as I was doing so, the little light lit up over my head, like in the comics. I pulled over into a parking lot and raised the hood. The distributor wasn't hard to find since all the spark plug wires led to it, and a quick twist confirmed what I suspected. The distributor clamp was loose and the distributor had slipped enough to retard the spark to an almost fatal degree. I gave it a crank in the direction that sped the idle up and closed the hood. It ran just swell and had plenty of power. Well, plenty of power for 57.3 cubic inches. I stopped again, returned the distributor to where it had been and limped back to the car dealer's lot.

81

I bought the car at an agreed $100 cash and the remainder of the money in two days. Back at my desk and very pleased with the car deal (which is a rare feeling to say the least), I got a frantic call from the lobby. The receptionist said there was a crazy man in a double-breasted suit with mussed-up hair and panic in his face, insisting that he must see me immediately. I went to the lobby, receipt for the car sale in my hand. Double-breasted salesman's problem was simple. I had forgotten to actually give him the $100 down payment and he'd forgotten to ask for it. My receipt notwithstanding, he insisted that I owed him a hundred bucks. To his surprise, I agreed and peeled off the bills and gave them to him. He wilted visibly, thanked me profusely, and left, a man still employed as a car salesman. He had violated the primary rule of used car sales—get the money first.

The biggest project our Houston group tackled was the CART Machine. CART was the acronym for Continuous Automatic Recording and Testing. It was an outgrowth of the CAT technology but was designed to do the testing required for transistor lifetime evaluation. Transistors were consigned to an oven and run electrically while at an elevated temperature. Periodically they would be removed from the 'burn-in" oven and tested to see if there was any degradation in performance. These tests were being done manually, but getting the data and keeping track of it was a nightmare, hence the CART Machine project.

It was like a CAT on steroids. Each transistor that went through the machine was automatically identified by notches milled in its carrier capsule, and the test data was no longer pass-fail, but real numbers. The numbers were transferred via an IBM cardpunch to, what else, IBM cards for computer input.

This giant leap in requirements required punching a seven-bit binary number for each of ten tests into a large CAT-like card carried along with the transistor and then transferring that data to the IBM card. On the same IBM card was the "Shingle" number, which was the carrier that fitted into the oven fixture, the "block" number, which was the individual transistor number, the transistor's product code, temperature code, date of testing and previous history. It was a big and complex set of equipments and it didn't work very well. And I'm really sorry to say that one of the major problems was a brain-child of mine, covered in more detail than you ever want to see, by

U.S. Patent #3,022,000, "Multiple Punching Machine For Paper Tape, Cards, Etc." It was filed July 2, 1957 when I was still thinking about the original CAT design, and was issued February 20, 1962. This semi-Rube Goldberg took seven pneumatic on-off signals and punched a seven-bit binary code in the passing paper card. Well, most of the time. It was a good thing we didn't try to use it on the CAT. It would have used up all nine lives.

The problems with the seven-bit punches were mostly overcome by careful adjustment, but the problems with another of my "good" ideas, namely the seven-bit pneumatic analog to digital converter, lingered. I won't attempt to describe how it worked, or more correctly, how it was supposed to work, but let me say that I took one of these dreadful A to Pneumatic D Converters and kept it in my bookcase for years. I would look at it periodically to remind me of some of my swell designs. Not.

The CART lingered for a year or two under the expert care and feeding of Roy Brink and then was scrapped. Not one of our better projects.

But we built a couple of dandies late in the history of our Houston group. One of the last things we designed and built was a pre-can transistor tester. This meant that the transistor was tested before it was sealed in its can, which goes along with the theory that you should get rid of bad things as early in the production line as you can.

The Mechanization Group in Dallas Semiconductor-Components Division was building a really large automated transistor factory called the Ultimation Project. I still have the custom-made lab coat with the Ultimation monogram on the lapel. It's now covered with various colors of house paint since Mary Ruth used it for a more basic purpose after I brought it home.

But the Ultimation manufacturing system presented us, the test equipment group, with a never-ending stream of uncanned transistors in little inch-square carrier blocks. We needed to build a machine to grab the block, make contact with the transistor leads, test it nine different ways, and then either pass it on to be canned or drop it into the great reject bucket in the sky. All of this at 7,000 transistors per hour, or almost two every second.

We didn't think we could make a herky-jerky start-and-stop machine like the CAT run at this speed without flying apart, so we decided on a continuous flow. We fed the blocks into slots in the edge of a continuously rotating wheel about two feet in diameter. It was tipped up at a rakish angle and ran on a Nash Rambler front wheel spindle and bearings. Contacts were cammed out to grab the leads and slip rings connected the transistor to nine separate test stations, one after the other. Any failure tripped a mechanical toggle on the carrier that diverted the transistor to the trash bin. Aw, too bad. Otherwise, it passed and was sent down a track for the next process step. Yea!

It ran wondrously well, as silent and smooth as the CAT was noisy and spastic, and it could crank out an incredible number of transistors in a day's time. It was a really nice piece of mechanical engineering, primarily by the biggest troublemaker in the group, Burlie Bowen. Burlie was an old classmate of mine from Rice. He was the one, who, along with his roommate, had decided to see how many times they could listen to the record *Gloomy Sunday* before they went crazy. It drove the freshman in the room next door crazy instead.

At the same time as the pre-can test set was being built, we were hard at work designing and building an array of automatic test parameter distribution plotters. Try saying that fast. They were to be connected to the test results of the pre-can stations for added information on how the transistor manufacturing line was running. Without going into mind-numbing detail let's just say that these plotters generated bar graphs, or histograms, of the transistor performance. A bank of nine plotters was built, one for each of the nine pre-can tests. The transistor process engineers could tell at a glance if their production processes were running properly. In the past, by the time the hand-logged data got to the process engineer and he discovered that his production line had gotten a bad batch of chemicals, for example, thousands of bad transistors might have been built. This gadget gave real-time process quality feedback to the engineer in charge. They liked it a lot and so did we.

The basic curve plotter became Patent No. 3,101,555, with Harold Armand, William "Bill" Gilchrist, Perry Westmoreland, and myself as coinventors. I wrote up the patent disclosure and submitted it to Jim Dixon, our assigned patent attorney in the TI legal office. Jim called me and said I needed to come over to his office and "flesh it out a little more," I dropped by and he asked me to sit down and take this Dictaphone and get started. He told me to think of every possible way to make this automatic bar chart move the bar up incrementally and please speak plainly into the microphone. He, personally, had to go to a meeting, but I was to take my time and give him more material to work with.

Naturally, when I ran out of real things to talk about I got silly. I ended by verbally envisioning an inverted tube full of water with electrodes at the bottom, and each desired increment of increase would be a pulse of electricity into the electrodes. The resulting bubble of hydrogen and oxygen would then float to the top of the tube and the height of the gas column would be the sum of the increments. The fact that it worked upside down was immaterial. And then came the best part—to reset the column back to zero, a spark plug in the top of this tube would be sparked. This would then detonate the highly explosive gas mixture, thereby resetting the indicator. I knew Jim would find this incredibly amusing. Well, he either didn't find it funny or more likely decided to get even with me. I refer you to Figure 10 of the patent. The spark plug is Reference Item 198.

My time in the Houston plant was winding down in the summer of 1960. Mary Ruth and I wanted to move back to Dallas. and I had talked with Nygaard about rejoining his group. And so. the 20th of August 1960 was my last day of work in the Geosciences and Instrumentation Division of Texas Instruments. Houston. Texas.

A few weeks before I was to leave. TI had appointed a friend of mine as Safety Coordinator for the Houston plant. Roger was a really nice guy who worked in the purchasing department and was very good at his purchasing job. What he did to deserve the honor of Safety Coordinator I don't know. but he wasn't near the bad-ass needed to enforce safety rules in a manufacturing plant. And of course, when he surveyed our work area he found a stack of aluminum boxes that were "precariously," the word he used in his report, piled on top of a cabinet. He was almost embarrassed to point this out to me. and begged my forgiveness for finding a flaw in our territory. I grudgingly moved the boxes to a safer location. pointing out to him that they would have scarcely caused a major scalp wound and certainly no more than two or three broken bones if they had fallen off the cabinet onto a passer-by.

He then made the tactical error of telling us that he would be back the next week to see that we were in compliance. We decided he needed something to really get excited about, not just a lousy fifty pounds of teetering sharp-cornered boxes.

Our plan was formed within minutes. To start things off, Joe Rich drilled a small hole through the bottom of a Pyrex baking dish with a carbide drill bit, which was no mean trick. The dish was then securely bolted through this hole to the top of a supply cabinet in a really scary-looking position, hanging over the side and appearing just about ready to fall off. A little modeling clay on the screw head sealed the hole in the dish bottom, and a quart of tap water colored with a dash of leftover coffee was poured into it. We completed the scenario with a piece of masking tape on the side of the baking dish marked "BATTERY ACID."

It worked perfectly. Roger was brought to the verge of a heart attack and I was told he never bothered the group again. He was poorly suited for the job anyway.

Chapter Five

Late August found us on the road again—family, furniture and all—headed back to Dallas. This time to temporary lodging in a small rent house on Lovers Lane. It was convenient to North Central Expressway, the four-lane pipeline to the new TI Expressway site, and close enough to the excellent University Park grade school for Number One Daughter Beverly to walk to her second-grade classes. Our plan was to find a house to buy in a month or two, and then really settle in. This turned out to be rather optimistic.

We looked a long time for a house that we both liked and could afford and found nothing that rang our collective bells. So the temporary rental residence lasted almost one year to the day when we moved into the house we had built in Lake Highlands. The moving boxes that had been stored in our rent house garage for a year were finally opened, and one mystery was solved. The missing sugar bowl, complete with crusty sugar and spoon, was packed with the last twelve volumes of the *Encyclopedia Britannica*.

The new Texas Instruments Expressway Site began with the building of the Semiconductor Building, a slightly exotic-looking saw-tooth roof structure. It was designed specifically to house TI's transistor manufacturing facilities. A noteworthy feature of the building was the "space frame," which was described as a "basement between the floors." The design allowed the manufacturing areas to be on the upper floor, which were huge wide-open square bays, 60-something feet across as I recall. Each square bay was topped with a single pre-stressed concrete roof unit in the form of a hyperbolic paraboloid. This was really neat looking and from its geometric shape, immensely strong. A trial roof unit was built alongside the building early in the construction phase and weighted with sandbags to destruction. Mathematics is one thing and testing the finished product to see if it has a tendency to fall on your head quite another. It passed.

The space frame was a heavily braced area between the bottom floor office space and the upper floor manufacturing bays. From experience in the Lemmon Avenue building, it was much simpler to supply all the plumbing and wiring that is necessary for manufacturing equipment from below instead of above. Drains especially work better that way. The external equipment, like vacuum pumps and water purification systems could also be neatly installed in the space frame. It was a really good idea. Leaks into the office areas below were rare.

Since the CAT machines and everything else we were building had been installed there, I was quite familiar with the new Semiconductor building. But Jim Nygaard's Dallas group first moved from the Lemmon Avenue plant into an old farmhouse on the Expressway site, and not into the new building. The 300-acre site had a number of usable farmhouses on it, and these were put to good use for small groups. One of the larger farmhouses became the Electrical Model Shop which built electrical things for the Semiconductor Division. Nygaard was always jealous of their palatial farmhouse. It even had a wine cellar.

Jim's farmhouse was a little on the shabby side, but it was convenient to the south end of the Semiconductor Building. The farmhouse parking and the path to the S/C building were gravel and quite a mess after a heavy rain, but this minor inconvenience was outweighed by being away from the early hassle of a new building. The main disadvantage was that Mark Shepherd knew where they were, and bugging Nygaard was one of his favorite indoor sports. He would drop by unannounced and sneak into the side door of the farmhouse and then take great delight in kicking the door to Jim's "office" almost off its hinges. The first time I was there and he did it I jumped out of my skin. It sounded like a gunshot. Followed, of course, by Mark's big laugh.

Various groups were moving into the SC building even as the construction was being completed. One of the early movers-in was Lee Kitchens, an electrical engineer from SMU and a manager of a transistor design group. Lee was spotted carrying his own tool box into the plant and was accosted by a covey of Union workers telling him that he couldn't do that—anything being brought into the building had to be moved by Union movers, so put it the hell down.

Lee, who is a bit less than four feet tall and with an abundance of attitude that more than made up for his altitude, told the workmen what they could do, both individually and collectively, with their suggestion. He then proceeded to carry his toolbox to his office. The workmen walked off the job. I don't remember hearing that they were missed.

I worked with Lee off and on for many years at TI. Having a meeting with Lee in his office required sitting in a short-legged chair with my knees slightly higher than his backup table. He had adjusted his office furniture to the proper scale of comfort with a hacksaw. Lee was a man of action. He was later manager of several of the TI calculator programs and ended his long career with five years as a TI-sponsored Visiting Industry Professor at Texas Tech University. He stayed two more years as a full professor before retirement. He now stays active in the Little People of America, serving as Vice President of Membership at last report, and has recently declined to run for Mayor of his West Texas town after a ten-year hitch. And don't forget his driving cross-country to attend the annual Rolls Royce owners' convention. Other than that he just sits around being retired.

But I best remember Lee for an incident that occurred when we were working on something in the space frame near noon. We decided to break for lunch, and Lee said he knew a shortcut to the cafeteria through the space frame. I told him to lead the way.

Lee headed out through the maze of bracing, plumbing and miscellaneous equipment in the space frame. It looked like the tank traps on the Maginot Line. I ducked, twisted. squirmed, climbed around, over and under, and otherwise made my way in the general direction Lee had gone, losing ground with every step. Lee, walking at a steady pace, arrived at the secret stairway to the cafeteria five minutes before I did, looking as natty as usual. I, however, looked like I'd been through a hard day in the garbage dump. I was dirty. dusty, hair askew and had torn my sleeve on a sharp corner. The good news was that I wasn't bleeding. I think he meant it was a "short" cut, not a "shortcut."

Of course, it worked both ways in the Semiconductor Division. Lee returned from an out-of-town trip to find that the office wall panel containing his door had been replaced with a solid panel. I remember seeing Lee on top of a stepladder. cigar clamped in his teeth, looking down into his perfectly functional but doorless office. My educated guess was that Mark Shepherd was responsible. No one else had the power to get the TI Facilities people to do something as energy-intensive as changing out a wall panel for a joke, especially twice.

But, remarkably enough, I was working hard at this time. We were working on upgrading the CAT machines, as the world of transistors was changing daily. We were scrambling to keep our testing technology up with the transistor technology. Nygaard's group, including me, was now in the main SC building and we were going ninety miles an hour.

In the fall of 1960, IBM had placed a big order for special germanium mesa computer transistors that were really tough to produce. Jim Lineback, head process engineer, and others struggled to find a recipe for building them that worked, and they weren't successful. The yield—the percent of transistors that were good out of the number that were started—was pitiful, being a few percent at best. IBM was beating on TI to ship thousands of units, not hundreds, and they absolutely had to have 100,000 units shipped to their plant by December 31 or it was the end of the world. But, if we could do it, they would accept and pay for an additional 100,000. Two hundred thousand transistors would not only make IBM really happy, but it would make TI a lot of money.

Lineback and his process engineers had one slim shot at fixing the yield problem. An obscure production process that had looked promising in trials but had not been fully evaluated was the only hope. But it was vastly different and difficult to implement. If the production line switched to it and bombed, the line couldn't switch back to the original process in time to get any quantity of units out to IBM by the end of the year. You could either struggle along trying to improve a cranky, low-yield process, or you could bet your ass on an untried one. Jim Lineback and his engineers had a meeting and, as one, bet their ass, and TI's, on the new and unproven process. Gulp.

And it worked beyond their wildest dreams. The production line yields jumped up phenomenally and good transistors began pouring out of the CATs on their way to IBM.

The big sigh of relief lasted about ten minutes as Lineback went to Shepherd with the good news and told him they were going to go for the 200,000 mark by year-end or bust their butts trying. Mark never batted an eye at this wonderful statement, and replied, "Well, *I'd* like to see three hundred thousand transistors!" From a couple of weeks before when the chance was slim to none of making

90

even close to 100,000 to a goal of 200,000 seemed like a gutsy call
but, as usual, Mark wanted even more.

We all scattered out into the manufacturing area to speed up
everything we could get our hands on. I worked on the can welders
and added a gadget that automatically kicked the finished unit out of
the electrode. This left the operator's hands free to load the next
unit, which sped it up by a third. We worked literally night and day,
and on the thirty-first of December, New Year's Eve to some lucky
people, we officially shipped the last of 200,000 transistors to IBM.
Hot damn!

In the midst of this "fire drill," the words commonly used to
describe a hectic, balls-to-the-wall, no-holds-barred effort, I closed
my desk about two in the morning and was heading home for a few
hours sleep. But as I walked into the hall, I noticed that the interior
patio area of the SC building was filled with sleeping sparrows.
Every patio tree was perched to capacity with snoozing birds, safe
from the winter winds. I suddenly had a great idea. I would sneak
out into the patio, catch a sleeping sparrow and put it in Nygaard's
desk drawer. That would be great in the morning. He would open his
desk drawer and a bird would fly out, probably causing at least a
mild heart attack. Perfect!

I quietly slipped through the door and into the patio carrying
a paper sack. What could be simpler than to catch a sleeping sparrow
and stick it into a bag? Well, as it turned out, practically anything.
You just *think* they're asleep. As if by magic, there was never a
sparrow closer than two feet to my hand, no matter what I did. They
flowed like water away from me, seemingly without even waking
up. I was greatly disappointed by this failure of another really good
idea. And I'm sure Jim would have been disappointed too, had he
known. A little excitement in his life would have been good.

But Jim had an even better idea for stirring things up. He and
the process engineers were mildly annoyed by Mark Shepherd's
request to see 300,000 transistors instead of telling them what a
wonderful job they were doing on the 200,000. So after the
December panic was over, Jim and the gang decided Mark ought to
see 300,000 transistors. All at once and on the floor of his office.

Because of the lousy original process, there was no shortage
of rejected transistors in the stock room, and Jim talked the clerk out

of 300,000. Boxes and boxes of paper bags filled with scrap transistors began arriving in our area. Fortunately, the transistors were quite small, like a quarter of an inch long with half-inch wire leads.

Bobby Howell, chief do-anything guy for Nygaard, was dispatched to find a large industrial-strength wheelbarrow, and Jim said he didn't want to know where it came from, just get one. So Bobby, as usual, was back within five minutes with the biggest wheelbarrow I'd ever seen. He said the gardener only chased him as far as the door and then gave up.

We began ripping open sacks and dumping the transistors into the wheelbarrow. A running total was kept from the quantities marked on the bags and, after an hour or so, we had achieved 300,000 transistors in the wheelbarrow. It was just right—the wheelbarrow was full to overflowing. I have never seen so many transistors before or since.

Jim Lineback, the man primarily responsible for the success, was the designated driver. He was flanked by Don Benefiel, Jim Nygaard and all the rest of us workers as he grunted the wheelbarrow down the long hall to Mark Shepherd's office at the far end of the building.

Lineback steered the load into Mark's office unannounced and the rest of us crowded in behind him. "You wanted to see three hundred thousand IBM units? Well, here they are!" said Jim as he upended the wheelbarrow on Mark's floor. Mark took it good-naturedly, possibly because he was outnumbered twenty to one. By sheer coincidence, the photographer from the TI S-C Bulletin just happened to be with us and captured the moment on film. It made an inspiring article in the January 31, 1961 issue.

If you'll jump ahead with me ten years to 1971, the same scenario was played out again in Mark's office by Jim Nygaard and a few cronies. Except this time it was with a toy wheelbarrow about a foot long and a handful of silicon wafers with integrated circuits on them. These few wafers contained more than 300,000 transistors. Now, of course, a single memory chip has millions of transistors and would be lost as a speck on Mark's floor. What a business.

1961 is known in my family as The Year That Everything Happened. It began in July with my father's death from a heart attack, followed in August by our family moving into our new house in Lake Highlands. The birth of our son David in October was next, with the year ending sadly with the sudden and untimely death of Mary Ruth's mother just before Christmas. It was a year of big ups and downs. But I also remember 1961 as the year we designed and built the Super CAT at TI.

The CAT machine was the de facto standard transistor tester at TI by this time, but it was becoming long in the tooth. The transistor technology was advancing at such a rate that it took a substantial effort to rig the CAT machines to do the advanced types of testing that were required. It just wasn't built for the new high-frequency testing, having been designed for DC tests. So we muddled along, spending a lot of our effort jury-rigging the CATs while we thought about the next generation test system.

Jim Nygaard's design-and-build team was now Bobby Howell, J.C. Baggett, Bob Chanslor, Jim Ricks, Lee Blanton, Jim Anderson, Harry Waugh, Troy Moore and me. It was just the right mix of brains, brawn and irreverence. We had no doubt that we could build the best transistor test machine in the world. We'd already done it once.

We had several all-hands meetings and discussed the shortcomings of the CAT that we had to overcome, and a wish list of things we'd like to see on a new one. We all agreed on what needed to be done. It's just that no one had any good ideas on how to do it. But Lord knows, we tried.

The memory of that time is still fresh. We all kind of wandered around in a daze for nearly two weeks, thinking about how to build the next generation machine. It was going to be called the Super CAT. We had the name but no design.

Every so often, someone would buttonhole another member of the team and say something like, "Hey! What if we took two plates about a foot square and drilled a big hole.... Wait a minute, that wouldn't work because of the connector... never mind." And the milling around would continue. "But if I screwed another piece on at right angles I could mount the connector off to the side and... no, that would interfere with the punch."

And then one fine day, about the time I thought we'd lost our touch, Jim Ricks stopped me and said, "What if we used half-inch separator plates between the test stations, then we could mount the test can on one end and the punch on the other." And I jumped in,

"Yeah! Then the track would go on top between the two with the card sticking out the back of the block instead of the front." And then someone else who had heard us and wandered up and interrupted, "Yeah! And then the block could have a notched base for the pusher pawls...." Jim Ricks had triggered the avalanche and we were suddenly off and running in a design frenzy. Others joined in and within about thirty minutes the basic Super CAT framework had been identified. I honestly don't remember another design effort quite like that one.

The final Super CAT was everything we'd hoped for. It was incredibly flexible and could be upgraded almost indefinitely. It ran at 4000 units per hour, or more than twice as fast as even the souped-up versions of the CAT. It was a GSB. And the Super CATs lasted until the small-signal transistor market was eaten up by integrated circuits and went away. *Sic transit* and all that...

We set to work building a prototype of the Super CAT, and now that Nygaard's team had Jim Anderson, an honest-to-God draftsman, we could even make drawings of the parts. What an improvement over the back-of-the-envelope sketches we'd been using. Meanwhile, Bob Chanslor, J.C. Baggett and Troy Moore were making knee-deep piles of chips in Nygaard's little machine shop off the office area. Parts were coming together and it was beginning to look like a machine.

One of the last units we designed was the Sort Station. Like the CAT, it "read" the punched paper card (still two-ply blue index, just bigger) carried along with the transistor being tested. We had a fine looking design mocked up, but we faced a tactical problem. Jim Nygaard liked to get his two-cents worth into the design and, since he was the boss, we let him. But this time, our design was absolutely perfect. It was functional, easy to build, the sort masks could be changed by the line operators without tools, and it just plain looked good. We liked it a lot. So we quickly built the worst-looking lamp housing in the world, installed it in place of our gorgeous one, and called Nygaard.

Nygaard's critique was simple. "I like it fine but that's the ugliest goddamn lamp housing I've ever seen. Fix it and then let's build some." It was really easy to fix.

This group of Jim's was a dandy. Lee Blanton, the quiet brains behind everything electronic, had been discovered by

Nygaard working in the TI Repair and Maintenance group. Nygaard correctly recognized him as an exceptional talent and got him moved into his group over the howls of the R&M manager. Lee was later elected a Senior Member of the Technical Staff of Texas Instruments in the first group so honored.

Jim Ricks, whiz-bang electrical technician and general wise guy, later went on to become the manager of the worldwide Field Service Group in Houston. His local claim to fame in Nygaard's group was his incredibly cluttered desk and hutch top. He resisted Nygaard's threats about cleaning it up by saying that he knew exactly where everything was. One day Nygaard called his hand. "Ricks, you have exactly one minute to hand me a 12AX7." Nygaard looked at his watch and said, "Go!" and Ricks went. He rooted through boxes and drawers with an ever-increasing intensity as the time ticked down. Parts began flying in the air from his desk. We retreated to a safe distance and watched with great interest. Just at the last instant, Ricks handed Jim the 12AX7 vacuum tube, with a "See? I told you I could find everything!" Of course, his desk and local area were a total shambles, and he had to clean it up anyway. Maybe that's what Nygaard had planned in the first place.

Bobby Howell was a terrific technician and Nygaard's Utility Infielder. He could figure out how to do or get anything. Once when Nygaard was going to make a point with someone about something, he sent Bobby out to bring back a dead crow. Seems Jim needed to serve it to someone on a platter. Bobby went home, got his shotgun, drove down to the Trinity River bottomlands, shot a crow and was back in a couple of hours. That's very fast for dead-crow delivery.

J.C. Baggett was one of the funnier guys in the group. He was built like a fireplug, and had a background in hot-rodding and could build most anything as long as he had a milling machine and a welding torch. The story goes that he was working on a car project in his garage at home with his oxy-acetylene torch turned up high as usual, the flame about a foot and half long, when his wife came out and asked him a question. J.C., virtually blind with his dark welding goggles on, turned around and said, "Yes, dear?" and cut the front out of his wife's nylon blouse with the flame from his torch. It didn't make a mark on her.

Troy Moore, the young punk kid, was a natural machinist and idea guy. He quietly worked with Bob Chanslor and J.C. Baggett on the mills and lathe. Or maybe he just couldn't get a word in edgewise. Troy continued his career at TI and is still working there as I write this. He and I worked together a long time. He helped me get a lot of raises over the years and is still one of my favorite guys.

Jim Anderson, artist, designer and draftsman, was a kind of misfit in our group. For one thing, he was polite. Also, he drew pictures of the machines on paper, which made him both a novelty and useful. I vividly remember him being stunned speechless when Nygaard gleefully found an error of eight feet in a thirty-foot, two and seven-eighths inches long Super CAT layout he had done. It held the record for the biggest identifiable mistake in our group for a long time and I've got a copy of the drawing to prove it.

Harry Waugh, soft-spoken Louisianan, had been a key player in the CAT development with his high-temperature oven designs. An LSU electrical engineer by degree, he could do mechanical, electrical, pneumatic, hydraulic, structural, aeronautical and about any other kind of design you could think of. If you needed help on a design, you called Harry. Every year I lost a dollar betting with Harry on the LSU-Rice football game. I was ever hopeful, and still am.

And then there was Bob Chanslor. Bob was special in lots of ways. TI originally hired him as a watchmaker to build tiny military products, but he was soon transferred to work with the early transistors. He was a fine machinist and watchmaker and one of the funniest fellows in TI. Bob spoke in "Chanslorisms," as they became known. In 1980, a list of these was published in honor of Bob's 30-year "badge party." Here are a few of my favorites:

> *"I'm so hungry I could lick the sweat off the kitchen winder."*

> *"He could wreck an anvil with a wet warshrag."*

> *"Boy! You sure hit the nail right on the thumb!"*

> *"What do you mean, build two alike? I can't build one alike!"*

> *"Mark Shepherd and I have a lot in common. We've both gone about as far as we can at TI."*

The list has 53 "sayings," but you get the idea. Bob could carry on a considerable conversation without using anything else.

In addition to his talents at TI, he was also a fine fiddle player. Not a violin, as he used to point out, but a fiddle. He could play breakdowns, hoe-downs, and any C&W you ever heard. During the 1930s, he was the fiddler in a traveling western band.

There was a group of us that would get together for "pickin' sessions" at my house, or Chanslor's house, or somebody's house. Nygaard played the banjo, Bob the fiddle, Steve Karnavas the drums, Jack Morgan the trumpet, Dewey Parker the electric guitar, and I had my guitar and mandolin. The group varied from time to time but these were the main "pickers."

One day at work Bob asked me if I would be willing to play a noontime gig with the Wells Brothers Country Band, a group that Bob regularly played in. It was run by his brother-in-law, William Wells, to advertise his Wells Brothers Feed Store in Plano. They were scheduled to play for a Lion's Club luncheon in McKinney, just north of Dallas, the following week. Bob said they were only going to play a short four-piece set, and it was stuff I was familiar with. So I said yes. Then Bob told me, "Oh, yeah, you need to play the bass." Oh, swell. I'd never played the bass before. Bob assured me it was a lot like the guitar and would be a piece of cake for me. Not quite, Bob.

So I borrowed the band's Fender electric bass, and made a tape of the four songs on my guitar. Every night I practiced with the recording and tried bass patterns. The day came, and we played, and we got a free barbecue lunch for our efforts. I was pleased that they didn't throw food at me. And I discovered that I liked it, so I ended up buying an old stand-up bass and going in business for myself, so to speak.

I eventually played bass in the Wells Brothers Band for three or four years. It was fun playing with good musicians, especially Chanslor. A typical "gig" for us was an invitation to play at the Collin County Fair, in McKinney. A dozen or two miles north of Dallas, the county fair was an annual event and drew big crowds. We, the Wells Brothers Band, were playing on the back of a flatbed truck under a string of bare light bulbs and had drawn a nice crowd of admirers. Or so I thought. In the middle of an especially fine number, the public address system announced the beginning of the

hog judging in a nearby barn. Our crowd left us for hog judging.
Now that was *really* rude.

Bob, of course, was in the middle of all of the jokes and
pranks that went on in our TI group and finally decided it was his
turn to do one. He had seen a really large weather balloon in a
surplus catalog that could be his for a few dollars, so he ordered it.
He had no problem enlisting the aid of everyone in the group to get
this balloon into Jim Nygaard's office, fully inflated. It was Bob's
first practical joke on anybody, much less on the Boss, and he
wanted Nygaard to think it was really funny.

On the morning of the planned assault on Jim's office, we
observed that Jim was called to his boss's office because of some
problem. He would be gone plenty long, so we sprung into action.
The first snag occurred when Bob actually read the instructions on
the outside of the Really Large Balloon package—"Balloon must be
boiled in water for two minutes before inflating to soften the latex."
So we scrambled off to find a hot plate and a pan big enough for this
unscheduled operation. But soon the balloon was boiling away and
we were rigging the vacuum cleaner blower outlet to a long hose
strung over Nygaard's office wall.

The flaccid balloon, covered with wet talcum powder, was
connected to the inflation hose and the blower turned on. Like
magic, it rose slowly from a pile of limp tan folds of latex into a
living, breathing thing. It gradually expanded into an elephant-sized
object, and, as planned, filled up Jim's office. When it began to
pooch out of the door opening we shut down the blower and taped
the hose shut.

The timing was perfect. Here came Jim, royally pissed after
an ass-eating by his boss. Aggravating his bad humor, I found out
later, was the fact that it was for something I'd done.

He saw the balloon bulging out of his office door and
stopped. Then the phone on his desk began ringing. Jim never said a
word, but scowling, pushed by the balloon and disappeared into his
office. We thought it was going to be a lot funnier than this.

Jim finished his phone call, squeezed back out of his door
past the balloon and then stabbed it to death with a pair of scissors.
It expired with a great talcum-powdered *whoosh!* all over his office
furniture. There was not a hint smile from Nygaard. His dark blue
suit was now smeared with the wet talcum, which added a surreal

touch to the unfortunate tableaux. Chanslor was horrified by the ugly turn his joke had taken and was never quite the same afterwards. Jim finally thought it was funny about twenty years later.

Jim's problem was he had little practice being on the receiving end of the jokes. He was much more often the instigator. For example, he had gone into the accounting group to see about something and needed to sharpen his No. 2 yellow pencil. He was directed around and across and down some more to the only pencil sharpener in the whole area. Why was this, he asked. George Livings, the head of the group, replied that one sharpener was plenty and there was no need to spend good money for a redundant one.

A week later, a large box from an office supply store arrived in our area. That night, led by our fearless leader Nygaard, we installed about forty pencil sharpeners in George Livings's accounting department. The fun part was trying to find enough really weird places to mount them.

Jim's brain was always working. One day he noticed that the sign over the Work Simplification training room door was nicely lettered in Old English script. He thought that was the silliest thing he'd ever seen, since it took someone a lot of effort to letter the sign, and with a nice sans-serif font it could have been done in one-tenth the time. Jim lodged a formal complaint, exactly where I don't remember, and within a week a new and simplified sign was over the Work Simp door. No one was safe.

And of course, Nygaard was not safe from his bosses. I was with Jim up on the production floor one afternoon when a messenger came up and told Jim he was to go to Fred Stote's office immediately. There was a production problem that would require Jim to cancel his imminent vacation. With a serious face, Jim headed down to find Fred.

Fred Stote was the manager of the Semiconductor-Components Mechanization Department where most of the production equipment was designed. I was told that as Jim entered, he was met not only by Fred Stote but also by Mark Shepherd and Charley Clough (as in "rough"). He also noticed a birthday cake on Fred's backup table with "Happy Birthday Jim!" on it.

He was relieved to find the "fire drill" was only a ploy to stir him up and get him down pronto to a birthday party. And he was pleased that they remembered. He shouldn't have been. They closed

the office door and began singing *Happy Birthday* as Charley lit the
fuse on the side of the cake.

The story is that all four ended up seeking safety under the
table as Charley had overdone the fireworks. After the last
concussion had died away in the smoke-filled office, the table no
longer held a cake. Its remains were now plastered on the walls, the
ceiling and the desk of Fred Stote. Jim's birthday party was a great
success.

A draftsmen sitting outside the office during this "birthday
party" told me later that he was concerned when he saw Jim
Nygaard, frowning mightily, enter the office with Stote, Shepherd
and Clough. The door closed and then what sounded like gunfire
erupted from inside. He said he was sure they were having a fight to
the death over some transistor production problem and was greatly
relieved to find it was only an exploding birthday cake.

Jim Nygaard tried to keep us on the cutting edge, even before
that expression became popular. In 1961 he somehow convinced the
powers that be, and those that controlled the money, to let him get a
computer for his group. Why? Because he thought they needed to
find out what those new-fangled things were all about. Who knows?
They might be useful someday. And so without any real purpose for
it, we acquired an IBM 1620 computer, complete with paper tape
punch and reader. We didn't need any sissy IBM cards to run this
thing with. What's the matter with a good old-fashioned paper tape?
We soon found out.

So we all traipsed off to IBM computer school, taught on site
by IBM. Unfortunately, the regular IBM training instructor was
home sick, so the IBM repair guy was called to fill in for our class of
seven or eight engineers. It was dreadful. None of us had the least
idea of how a computer worked much less how to program one. I
remember distinctly the first words from our "instructor." They
were, and I quote, "The op code for 'ADD' is zero one." What the
hell was an "op code"? The class went downhill from there. That day
and the next were like a nightmare, where you're in school and you
don't understand a word the teacher is saying, and the big test is
coming up. I took notes like crazy and marked up the handouts,
hoping I could figure it out later.

The second evening at home, I was poring over the material
again, and something caught my eye in the IBM instruction material.

100

They had accidentally let something slip by that I understood. Ah ha! The op code is the first part of the memory word and the address is the last part! And then....!!! Again, if you had been in the room with me you would have seen the light bulb appear over my head, shining brightly. All the pieces I had been juggling in my overworked brain for the last two days fell into place with a big thud. So *that's* how computers work! If our so-called instructor had spent a single minute on the structure of the computer, it would have saved me a lot of grief. On the other hand, I certainly remembered it a lot better after digging it out by the roots. The class was a lot more fun the next day.

Jim decided that he needed a "computer person" to run this beauty, and was chatting with me about it one day. He said, "We need someone who likes to work by himself, and is smart and sneaky and devious to write the code for it." We looked at each other for a second and then said simultaneously, "McDonald!" Jim had described Mac McDonald perfectly. Mac had somehow veered off the TI track but was still living in Dallas and working for Continental Electronics building megawatt transmitters. He succumbed, as I so often did, to Nygaard's wiles and blandishments and hired back in to become Jim's Computer Guru. And, as we surmised, he was wonderfully suited to the job of writing computer code and finished out a long career at TI doing just that. Mostly because he was smart, sneaky and devious.

Lee Blanton also caught on to computer programming quickly, as he did everything. and was challenged by Nygaard to write a program that would calculate the spacing of frets on a banjo. This required some tricky mathematics, but Lee was up to it. I looked over Lee's shoulder and "helped" him, and Lee added a little surprise for Nygaard in the code.

Lee finally told Jim that the program was finished and if he wanted to design a banjo, come on by and he'd run it for him. As Jim and I walked over to where Lee sat at the IBM 1620 console. I whispered to him, "Put in a really big number for the banjo neck length and Lee's program will blow up!" Jim smiled sweetly and sat down beside Lee.

101

Lee said, "You've got to put in the length of the banjo neck and then it'll calculate the fret spacing for it. Just type in the length in inches."

Jim typed in "2-0-0" and hit the "Execute" button. The printer clattered out,

NYGAARD YOU CANT PLAY A BANJO THAT LONG

By the end of November 1961, our group had upgraded the Super CAT to the Instant Super CAT. This clever modification to the Super CAT eliminated most of the custom-built test cans and replaced them with standard, albeit more expensive, "programmable" test cans. All you needed was an easily patched-up printed circuit board about the size of a piece of tablet paper poked into the top of it and you were in business. Instead of a $50 or $100 test can for a new product specification, the printed circuit board could be made and programmed for $15 to $20, with the time reduced to about an hour. Plus, a technician could change all 30 or so test stations on a Super CAT from one type of transistor testing to another in less than fifteen minutes. He just walked down the line with a pushcart full of boards and swapped them out. It was a neat and very efficient system, with, of course, Lee Blanton at the heart of it. Or should I say brains?

And the Instant Super CAT also had its moments of excitement. If the programming board was plugged into the test can backwards, which was supposedly impossible to do since it was mechanically keyed, it would short out the heavy-duty 300-volt power bus. You really had to work at it to overcome the key, but when you did you were rewarded by a great "BANG!" and a flash of fire followed by a cloud of smoke from the partially vaporized circuit board. People rarely did this twice.

But about this time, yet another transistor disaster began to rear its ugly head. IBM was buying our transistors like crazy for their new computers but they were beginning to have more and more transistor failures. This came to a climax when an IBM vice president was inspecting a computer mainframe on their checkout floor in Poughkeepsie, New York. He pulled out his gold-filled Cross ballpoint pen, probably with an IBM logo on it, and ran it down a row of operating transistors, rapping each one in turn like a kid with a stick on a picket fence. The computer quit on the spot. He said in a manner that only vice-presidents can get away with, "I want TI to whack every transistor they build for us with this pen before they test it. And then, if it passes that, we'll buy it."

Would you believe we began by measuring the size and weight of a Cross pen and building a machine that would duplicate a vice-president hitting a transistor with it? Of course you would. Our IBM business depended on it. When IBM said "frog," we didn't ask how high to jump, we just jumped as high as we could. Hence, the origin of the infamous Tink-Tink Machine.

A cam-operated metal rod, closely resembling a Cross pen, would drop on the unsuspecting transistor that was held down on an anvil, kind of like Marie Antoinette's head on an automated chopping block. And sure enough, some of the transistors would be destroyed electrically after being "tink-tinked." The rest were shipped to IBM. The reliability improved marginally but the problem really hadn't been fixed, regardless of the man with the Cross pen.

A high-powered mathematical analysis by Dr. Warren Rice, our frequent savior at Arizona State University, showed that the Tink-Tink's "G-forces" at the transistor element, and at the tiny wires coming from it, varied all over the map. It was not a good method of stressing the element to cause failures. We quickly proved this by rerunning good units, that had already passed the Tink-Tink Machine, and having even more transistors fail. In fact, if you were patient, you could tink-tink any transistor to death. Big time bummer.

So Warren Rice and Jim Nygaard decided that the only way to properly stress the units in order to fail weak ones was to put them through a 38,000-G centrifuge test. This sounded good except we didn't have any centrifuges. Yet.

Warren, Jim, and Tommy Chaddick, manager of the big TI Semiconductor machine shop, sat down in Jim's office and designed an air-turbine-powered centrifuge. The design was completed quickly and Tommy set his shop to work and built several of these untried devices in just a day or two. They were installed in an unused farmhouse on the edge of the TI property. Sand-filled walls were built to stop fragments in case the unthinkable happened, like a wheel full of transistors exploding from the centrifugal force. A huge diesel-powered air compressor was rented that looked big enough to inflate every car tire in Texas at the same time. And these turbines used all the air the compressor could supply. They were powerful turbines and were able to accelerate the heavy wheel up to 15,000 rpm in less than a minute, but efficient they weren't. Efficiency of operation wasn't on our list of requirements. Getting our collective asses out of serious trouble with IBM was, however. At the moment, it was the only thing on the list of requirements.

The air turbine drove a vertical shaft on which the operator would load a thick, round aluminum plate about a foot in diameter. This plate held a hundred transistors in pockets milled in the edge. A cover kept the units from flying out during the spin.

After loading the wheel on the turbine shaft, the operator would retire to a safe distance, behind more safety shields, and open the air valve to the turbine. The air turbine would begin to wind up with a rising howl ending in a shriek that could be heard for a mile. The compressed air would then be switched to the "reverse" nozzles, and the turbine would brake itself to a stop.

A little sleuthing with a microscope had proved what Warren Rice's mathematics had indicated, namely that 38,000 Gs on the tiny gold wires would take a poorly-fastened wire off of its intended contact and straighten it out completely. No more partial connections that might pass a test and then fail if hit with a gold-filled Cross pen. The centrifuge stress test worked perfectly and the transistors shipped to IBM suddenly made a step-function improvement in computer reliability. The only problem reported by the centrifuge operators in the farmhouse was that stray cats had been using the open-topped sand-filled walls for other than their intended purpose.

The Mechanization Group took over the design of the next generation of centrifuges since it was almighty unhandy to have to carry the transistors out to a remote farmhouse for testing. In addition, the cost of the air compressor rental was substantial. Mechanization designed a neat, completely self-contained, electrically-driven centrifuge. It had a power-driven clamshell lid that would close and hold the spinning disk in an explosion-proof steel chamber. Recognizing that this was a dangerous piece of equipment, they added lots of safety features. One was especially good. The lid could not be opened to expose the wheel unless the drive shaft was not turning. It had to be completely stopped before it would automatically open for unloading. What could go wrong?

Well, I'll tell you what could go wrong. The new improved centrifuges had been running successfully for months in the Semiconductor Building manufacturing area. Then one day an operator forgot to tighten the nut that held the disk on the drive shaft. All was well as the disk spun up to the required 15,000 rpm, but as it began to brake to a stop, the nut unscrewed and spun off the shaft with the whirling disk close behind it. The shaft, now with no inertial load, stopped quickly indeed and signaled the lid-lifter that all was well and it was safe to open up. The disk, unattached to the now-stopped shaft, was still rattling around inside the bulletproof chamber at not much less than 15,000 revs. The safety lid indeed lifted, and the whirling disk hopped out of the centrifuge like a thing alive and hit the floor. Still spinning at a tremendous rate, it skidded on the floor until it gained enough traction to fire itself across the room, bouncing off the walls as it went. It ended its flight imbedded in a supply cabinet with no injuries to the wide-eyed operators. The unthinkable had happened and the safety system failed to catch it.

Sometimes luck is better than skill when things go wrong, as witness the centrifuge accident. But another close call occurred one night in the middle of the second floor manufacturing area. I was up there working late, like at eleven o'clock, along with the full complement of employees on the night shift.

I heard the sharp bark of a furnace tube hydrogen explosion at the far south end of the open bays and turned to see where it was. There were some screams from the ladies and a few got up and moved away, but things calmed down in a couple of seconds. Then came a second explosion. And then a third and a fourth. Within a minute there was a wall of screaming women moving down the aisles in my direction, pursued by a continuing series of furnace explosions. Some were high-pitched pops, some medium-toned bangs, and a few were deep, resonant booms from the big chain conveyor furnaces. To add to the growing excitement, the larger furnaces were firing red-hot graphite carriers across the room. I climbed up on the table to avoid the rush.

By this time the line foremen had taken notice and had begun frantically valving nitrogen into the furnaces. They didn't know what was happening, but, in case of an explosion of any kind, it was standard procedure to shut off the hydrogen or forming gas, purge the furnaces with nitrogen, and run like hell. The mayhem stopped as

105

quickly as it started. In a few minutes. everyone drifted back to their work stations and began to clean up the wreckage. But no one was injured. Just shattered nerves, including mine. I never did like hydrogen.

The answer to the moving wave of furnace explosions became clear the next day. The automatic forming gas mixing valve had failed. Forming gas. which was commonly used in furnaces requiring a reducing atmosphere, was used instead of hydrogen because it wouldn't burn or explode. That is. it wouldn't if it was mixed in the right proportions. Instead of 10% hydrogen mixed with 90% nitrogen. the valve had shut off the nitrogen completely and 100% explosive hydrogen was piped into furnaces designed for the non-explosive forming gas. Since the forming gas line fed the manufacturing area from the south end. the furnaces exploded in turn as the hydrogen progressed up the supply pipeline. So much for fail-unsafe automated gas mixing valves.

This was about the time that the very first "cleanroom" was built in the Semiconductor Building. The new-fangled cleanroom was needed to process the other new-fangled gadget. the integrated circuit. The complexity of the integrated circuit was much less forgiving of a speck of dirt than a transistor, and so the first cleanroom was built at TI.

The concept was simple. Special "HEPA." or High Efficiency Particle Air filters were used to clean the air that was distributed in the room. Most of the room air was then recirculated to join a small amount of incoming air and be cycled through the HEPAs continuously. The early cleanrooms would have about one-thousandth as many particles floating around in the air as the ordinary space in a manufacturing facility.

But. as with all new and untried things. problems arose. The first problem. as Jim Nygaard pointed out. was that the engineer in charge of the suck in the cleanroom was ahead of the guy in charge of the blow. The room was operating at a negative pressure with respect to the rest of the Semiconductor Building and consequently sucking in dirty air through every door and leaky wall joint. Major adjustments of the in and out airflows corrected this so that the room finally and correctly oozed air out into the surrounding areas and stayed clean.

The second problem was short-lived and related to me by Jack Kilby, a reliable witness. Boyd Cornelison caused it. Boyd, suspicious of integrated circuits in general, probably because it looked like they were going to someday supplant his beloved transistors, took a tour of the new cleanroom to see what the competition, so to speak, was doing. He walked into the cleanroom unsmocked and smoking a cigar. When chided by the line foreman for this gross breach of cleanroom etiquette, Boyd replied, "Well, it can't be much of a process if it can't stand a little cigar smoke!"

But this humdrum life in the world of transistors had about worn me out and I began to think about doing something different. Maybe a job with fewer hours per day would be good? I started poking around TI and came across Science Services. It looked like an interesting Division that did work for both the government and Geophysical Service. They were big in oceanographic and seismic equipment, which was about as different from building transistors as you could get. I decided to check out the opportunities, and before I knew it, an opportunity had turned into a job offer. I left my beloved transistors to go to work for E. B. "Eddie" Neitzel in Science Services Division on the 7th of May 1962.

Chapter Six

Going to work in the Science Services Division of Texas Instruments was like starting over in a new company. It was located in Dallas on the ninth and tenth floors of the Frito-Lay Building at the corner of Harry Hines Boulevard and Mockingbird Lane. A very non-industrial looking place from the outside, with shops and restaurants on the ground floor, but up where I was going to work it looked about like any design and development lab. There were plenty of workbenches and hot soldering irons. It's just that you got to work by riding an elevator. Cool.

Eddie Neitzel was my new boss and he was as calm and unflappable as Nygaard was hyperactive and intense. I had no problem with that. None at all. I had promised myself a change and I got one. Another big change was the civilized environment of an office building and seeing other tenants wearing coats and ties to work.

I had been hired to work on an interesting project that was a portion of an Air Force contract known as Project VELA UNIFORM. As part of the United States interest in the state of Russian nuclear weaponry development, an effort was being made to monitor the earth's crust for seismic signals that might be from Russian nuclear test shots. And the way you monitor the earth's crust for movements is to use a seismograph, like an earthquake seismograph. These gadgets date back a long time, and from a purely mechanical beginning now sported electrical transducers and output signals that could be sent to remote chart recorders for display. But, and this was the big catch, the end product of an earthquake or a nuclear blast was a wiggly line on a piece of paper. This narrowed the analysis tools down to one—the eyeball of an expert seismologist.

And even an expert seismologist had a hard time sorting out a nuke wiggle from a weak and localized earthquake wiggle. What they really needed to do was put the wiggles into a computer and then, with techniques developed in seismic oil exploration, let the computer sort out the characteristics that could identify the source of the signal with a fair degree of accuracy. All they needed now was a seismograph that put out the kind of signals that could be run through a computer. Enter Ed Millis, Stan Lehnhardt and Walter Ford. We three were the engineering team that was to develop a seismograph system that could talk in a language computers could understand—a digital seismograph system.

I met Walter and Stan along with others on my first day. One of the others was Tom McCullough who later joined our team as a computer expert, among other things. Tom was probably the brightest engineer I ever had the pleasure of working with, and I worked with some sharp cookies. He was a jack-of-all-trades and master of all. He could do just about anything.

A few weeks after I had met Tom I walked into his office and noticed that there was a crackling noise under my shoes. I guess I looked puzzled by it, because Tom volunteered, "That's just the remains of some ammonium tri-iodide that I mixed up. Pay no attention to it." My ears perked up, since I knew that ammonium tri-iodide was the notorious "explode when tickled with a feather" compound they used to demonstrate in chemistry class. I inquired further of Tom, as to why the Sam Hill he was mixing it up in his office?

"Well," said Tom, "I've been having trouble with the night cleaning lady messing with stuff on my desk, so I made some tri-iodide. I was in the process of painting it on the top of my desk so it'd blow her damn duster to bits, but it went off in the beaker and ended up all over the room." Oh, swell. Tom's office had an explosive floor, in addition to other things. "Yeah, it's pretty much on everything." He picked up a memo from his desk to demonstrate, and it sounded like Rice Krispies when he waved it at me.

I nodded knowingly and left, trying to walk in my same footsteps. I'd forgotten what I went into his office for in the first place. He was not only smart but interesting.

My job was to design the transducer and the mechanical parts of the seismometers for our new digital seismograph system. The transducer measured the movement of the earth and sent out the digital signal. To properly monitor the earth required three seismometers: one to measure the motion in a north-south direction, another east-west, and a third up and down. All three could use the same transducer, so I set to work on its design first.

109

I decided on a system of high-frequency oscillators that would be controlled by the position of the seismic mass. If the mass moved one way, the frequency would go up; the other way, down. I sent this variable frequency signal to Stan Lehnhardt to take care of. He put the cycles into a counter and counted them for a fortieth of a second. The number then represented the position of the mass, or approximately, the position of the earth, which is what we wanted to measure in the first place. And this was a number that a computer could swallow without gagging. Not too complicated, actually, and it worked just fine.

The main technical problem that I faced was trying to make an oscillator work at 450 MHz with the available transistors. Nowadays, just about any 98¢ transistor hanging on the hooks at Radio Shack would work, but that was not the case in 1962. I had to have the oscillators run at a high frequency in order to get the sensitivity that I needed to meet the seismograph specifications. I talked with my smart friends in the Semiconductor Division and they came up with the 2N2415 germanium mesa transistor. It was state of the art and then some. The "then some' meant the damn things cost sixty bucks apiece. Just to give some scale to this price, this was when IBM was paying less than a buck for a top-notch computer transistor. But of course, they did buy a few more than I did. The good news was that I only needed six, which was two for each of the three seismometers. I ordered ten of them to be on the safe side, since I'd blown out a few transistors in breadboards before. I drove out to the Expressway Site to pick them up in person. To me, it was kind of like getting a shipment of diamonds. Those suckers were precious.

At the same time, I began working on the seismometers themselves. A pendulum was needed that would swing across and back in ten seconds. If the earth shook. the frame holding the pendulum's top support would move back and forth with the earth, but the chunk of iron that was the main weight of the pendulum would stand still, relatively speaking. because of its inertia. Then all you had to do was put something between the frame and the mass and measure the motion. Now you know how a seismometer works.

A simple calculation will show that a pendulum that takes ten seconds to swing one swing is going to be a long one. Longer even than Nygaard's banjo neck. Like 88 feet long. Packaging this pendulum in a tabletop box was going to be a challenge, no doubt about it. Except for Fred Romberg. Fred worked for Science Services and was from a scientific family. His father, Dr. Arnold Romberg, had his name on the LaCoste-Romberg seismometer and a line of gravity meters. The seismometer was a clever mechanical system utilizing a "zero-length spring," which I thought was really far out and *very* scientific sounding.

With the help of Fred, I designed a version of the Lacoste-Romberg seismometer that had a pendulum eight inches long that thought it was 88 feet long. It worked well and I was thrilled out of my gourd to find that in the official reports to the Air Force it was called the "Millis-Romberg" seismometer. I even got first billing!

I liked mechanisms like the LaCoste-Romberg pendulum a lot. If you tweaked the spring suspension screw on the pendulum you could make it think it was infinitely long with an infinite period. I really liked that kind of thing. Imagine! An infinitely long pendulum! And so I began to think, which happened intermittently. The LaCoste-Romberg invention had the zero-length spring in it, which, although it sounded keen, had problems with temperature and was difficult to make. I tried to imagine a way to do it without a spring. So one afternoon, after I'd eaten all my TI spinach for the day, I sat down at my desk with a nice clean sheet of paper and a pencil and treated myself to dessert. I was going to try to design a long-period seismograph pendulum that didn't have any springs in it. After a bit of head-scratching I came to the conclusion that all I needed to do was find some peculiar linkage that would swing a weight in a path that a long pendulum would take. For example, an arc with an 88-foot radius. There's the light bulb again. I recalled a strange linkage called a Scott-Russell straight-line mechanism. This clever and simple mechanism could force a point in space to go in a straight line and was used occasionally in machinery. Well, I thought, if I could fasten a big weight to that point, I already had an infinitely long pendulum, since a straight line was the arc of an infinite radius. And I just bet that if I boogered up the geometry of this Scott-Russell mechanism just a little, it would no longer be a straight line, but an arc of a really long radius. Just what I wanted!

Now to try it out on paper. Oops, no compass to draw arcs with. Quick, downstairs to an office supply store to buy the biggest and best compass they had. Money was no object when a brainstorm was at stake. Ten minutes of drawing lines and arcs and sure enough, it could be done. I later designed and built several models of seismometers and tilt-meters using this idea just for fun. My patent disclosures were rejected because of a notable lack of a commercial market for TI. But I didn't care, much, because it had been a lot of fun to invent.

I am unable to resist the "Gee Whiz!" stuff about the horizontal seismometer that we built for the VELA UNIFORM project. If you were a computer and looked at the signal coming from this seis, you could see when the earth moved eight nanoinches, which is eight billions of an inch. An even bigger Gee Whiz is what happens when you tilted the base of the horizontal seismometer. Since it thinks it's a pendulum 88 feet long, it acts like one, too. If you tilted a frame that was tall enough to hold an 88-foot pendulum even a little, the pendulum weight would swing sideways a lot. And that was true of this system. It was unbelievably sensitive to tilt.

If the base of our horizontal seismometer was tilted an angle of 10^{-7} radians, the sensor would register a motion of the pendulum mass of a bit over a tenth of a thousandth of an inch, which was 2.5% of the full range of motion it could detect. That amount of tilt is the same as lifting up one end of a ten-mile long plank by one-sixteenth of an inch. *Gee, Whiz!*

To jump ahead momentarily, this system was soon installed and running in a cave near Albuquerque. The cave was a man-made hole in the side of a solid granite Sandia Mountain. The floor of this granite hole was smoothed out with concrete, and the instruments placed on it. Whenever I walked near the horizontal seismometers, I could see the signals change as the pendulum swung over slightly. I was bending the solid granite mountain with my weight. Honest. Now *that's* sensitive to being tilted.

And to close the loop that a few technical people might find I've left open, yes I had to design an automatic system to continuously level the seis because of earth tides from the moon, temperature effects, and other drifty spooks. I controlled the current to a heater coil on a leveling leg that could lengthen as much as three ten-thousandths of an inch with five watts of heater power. That could tilt the seismometer base five seconds of arc, a huge amount of tilt for the system.

So we built the system and the first tests in the Exchange Bank building looked pretty good. The tenth floor of a building was a lousy place to test a seismograph. Actually it was an impossible place, so we moved it to a borrowed space in the basement. We could easily detect a freight train passing on a track a half mile away. But the basement was still much too noisy, earth-motion-wise, to really test it.

We contacted the US Coast and Geologic Survey people and arranged to borrow a seismograph test cave at their main laboratory just on the other side of the Sandia Laboratories near Albuquerque, New Mexico. By this time Tom McCullough had joined us because of his computer expertise, and had signed on to help me in the long series of tests scheduled at the Geologic Survey Lab. That was good news for me since I didn't know how to run the data-collection computer that had been attached to our seismic system. We packed the equipment in a rented truck, got a driver from TI, and flew out to Albuquerque to meet it.

.   Safely ensconced in a cheap motel with a swimming pool, we headed to the Sandia Base in our rent car and arranged to get a pass that would allow us to drive through the Sandia grounds without being machine-gunned to death. If you didn't know, Sandia is where a lot of nuclear weapons are stored and worked on. Strategic Air Command long-range bombers regularly shuttled in and out with their nukes. It wasn't a good idea to get cute or make any sudden moves when you were passing through the Sandia property. The perimeter fencing around the main nuclear weapons storage mountain was five separate fences; each separated by a dozen feet. Patrols along these fences were under instructions to shoot to kill anyone found inside even the first fence. The whole area was some serious shit.

We checked in with the U.S. Coast and Geologic Survey Laboratory and were issued our own cave. It was away from the main buildings and on the side of a mountain a short distance up a rocky road that ended at the door. The steel outside door opened into a rough blasted-out tunnel strung with a few bare light bulbs. This thirty-foot tunnel ended at another steel door that led into the main room.

The room we were to occupy for what turned out to be seven weeks was about twenty-five feet square. The walls and the ceiling were "as blasted," but the floor was smooth concrete. The ceiling had an added feature of wire netting fastened to it to catch falling bits of granite. The room was perfect for our testing. After our truck arrived, we had our equipment set up and running fine as frog's hair within a day or two.

Tom and I spent a lot of time watching the needles wiggle on the chart paper. The tests of the seismograph system weren't hard to do but took hours. Often the equipment would be set up in a specific configuration and run for a half a day or a day to collect data.

The weather in New Mexico was generally nice in the late summer of 1963. Afternoon mountain thunderstorms were the norm and a welcome diversion. Except for one afternoon. As Tom McCullough and I were driving back into Albuquerque, we were red flagged to a halt and then allowed to fall in line behind a convoy trucking a nuclear weapon to the air base. We stayed a respectful distance behind, as we were motioned to do, and drove at the fifteen miles per hour rate of the convoy. The nuke was carried beneath a lumber-carrier vehicle that straddled it like a long-legged animal would straddle its prey. The weapon was modestly draped to shield it from the view of passers-by, but it was remarkably large.

As we crept towards Albuquerque, an afternoon thunderstorm developed and had begun moving in our direction. The convoy halted momentarily and then did a slow U-turn, doubling back along the road. The first vehicle to reach us was equipped with serious-looking military personnel carrying automatic weapons. We were told to pull our car way off the road and into the ditch and stay there until the convoy has passed. I'm surprised they didn't tell us to hide our eyes and count to a hundred.

So we got as close a look as I care to get at a canvas-swathed nuke when the funny-looking carrier crept by a dozen feet away. Tom and I speculated as to what might happen if it were struck by lightning, and we decided it wasn't worth the effort to try to put some distance between us and the bomb. We couldn't get anywhere near far enough in fifteen minutes.

Tom's impatience with the perversity of inanimate objects finally reached the breaking point in our testing regimen. The expensive 14-track digital tape recorder that we were using to collect our test results had been getting crankier and crankier by the day. Recovering from parity errors began to exceed the useful running time, and Tom was pissed. He dug a screwdriver and a pair of pliers out of our toolbox and attacked the beast's recording head.

The instruction manual had pages of "do nots," mostly concerning messing with the 14-track recording head. This head, it seemed, could only be cleaned and adjusted in a sterile factory peopled by gray-haired scientists who had devoted their lives to such incredibly delicate operations as realigning 14-track recording heads. And then, only when surrounded by thousands of dollars of unbelievably precise fixtures and measuring instruments. Tom, armed with his screwdriver and pliers, was not impressed and proceeded to dismantle the head into all of its component parts, which were a lot.

He spit on his handkerchief and cleaned a few critical parts and then reassembled it. He realigned the head by squinting at it with one eye shut and tapping on various parts with the screwdriver handle. Within an hour from the beginning of his frontal attack, Tom had reinstalled the head on the tape deck. Of course, it worked perfectly and there were no more parity errors. I have no idea how he did it. It was just the way Tom operated.

An eerily similar event took place in our motel room a week later. I wandered into Tom's room after I noticed his door was open and found him playing with some odd bits of metal on his bed covers. In answer to my question, he said he was just damn tired of the lock on his door being hard to operate, and the motel clerk had just shrugged his shoulders about it. So he had removed the lock assembly from the door and dismantled it on his bed. He was holding his key in the lock cylinder and checking the length of the tumbler pins as he talked. "Just as I thought!" said Tom, "The damn thing's pinned wrong for this key!" And so Tom juggled pins around in his door lock, reassembled it, still using the bed covers as a workbench, and put it back in his door. Presto! No more key problems. Do people just naturally know how to do that? I think not.

We were able to spend some hours hiking in the surrounding mountains as our instruments gathered data automatically. The bright sun and blue sky combined with the cool invigorating air was a pleasant change from our usual summers in Dallas. And there were a number of abandoned mines in the area that attracted us like magnets.

One in particular looked interesting and inviting. The entrance tunnel went straight and level into the side of the mountain an unknown distance. The remains of a tram rail roadbed made an easy approach to it, so Tom and I decided to explore.

The noonday sun was bright and we strained to see what lay beyond the black opening. We stopped at the entrance and leaned to see, shielding our eyes from the glare. I didn't actually see anything, but a rattlesnake saw us, took offense, and started a loud dry rattle at what seemed like point-blank range. Tom and I, as one, rose vertically in the air without apparent effort, turned 180 degrees, and accelerated to just below the speed of sound. No brain was required to think about what to do. It was the old, "Feet, do your duty!"

A few days later and after the purchase of a flashlight, we tried again. This time, a barrage of rocks thrown from a safe distance preceded us to the entrance, but our friend of the first visit had sought a sunny spot elsewhere. We walked into the mine for hundreds of feet. It was clean, dry, and solid granite except for a thin vein of galena ore that the tunnel was following. A long way in we came across a ventilation shaft angling out the top of the tunnel, and a vertical shaft that appeared to go down to the center of the earth. It had the remnants of a rickety wooden ladder on its wall, disappearing into the blackness below. It was an altogether scary proposition. I am in awe of the brave men who worked in that mine.

Tom McCullough and I had drifted into a comfortable routine of sleeping late and working late. This was interrupted regularly by great helpings of New Mexico-style Mexican food that I love to this day. Occasionally we worked all night if we were making a long test run on the equipment.

On such a night as this, I was in the corner of our little granite cave practicing music on my alto recorder. For those of you not musically inclined, this is a wooden flute-like instrument that dates back to medieval times. Mine wasn't quite that old, but I had been diligently learning how to play it all year from the do-it-yourself Trapp Family Recorder book, and was making encouraging progress. For the first time in my life I could actually read music.

116

What a revelation! I could now turn black spots on a page into a tune!

It was one or two in the morning and, with my book of music propped up on the digital seis console, I once again started Bach's Brandenburg Concerto No. 2 in F Major. I was enamored with the idea that I, Ed Millis, might be able to play a piece of music that Bach had written. Far out, indeed! So, once more from the top...

I was in the middle of the second line when I was startled by the sudden bang of our inside door being slammed open, and jerked around to see what was happening. My astonished gaze was met by the equally astonished gaze of a Military Policeman with drawn revolver.

With a relieved smile, our visitor lowered his gun and said, "Boy, you guys scared the bejeesus out of me!" Out of *him?* He didn't know the half of it. There was some coffee on our hotplate, so we invited him to share his experience with us, as they say now.

It seemed like he was on his guard rounds checking the Geologic Survey property and saw a faint light on the side of the mountain. He cautiously approached it and he heard a terrible wailing coming from inside. He could not, he said, no matter how hard he tried, imagine anything that could make that kind of noise, and so he was prepared for battle. That kind of hurt my feelings since I thought I was beginning to sound pretty good. But in any case, we parted friends; all three of us immensely relieved that it had been a false alarm.

Our sluggardly habit of rising late in the morning was interrupted by the information that there would be an underground nuclear test shot at Jackass Flats, Utah, and it was imperative that we record the seismic shock wave from it. We had already caught a few earthquakes and we needed to record a nuclear blast for comparison.

The shot was to be detonated at 9:00 AM so Tom and I planned to get up in time for a quick cup of coffee, drive to the cave, turn on the digital seis and nab that sucker as the waves rolled through. What could go wrong? Does this sound familiar?

On the morning of the nuclear shot, we both overslept. After the proscribed door pounding and yelling, we threw our clothes on as we jumped in the car and sped for the cave. It was still some minutes before nine and we were going to give it a good try.

As we skidded around a turn and were accelerating down the straight, still a mile from the Geologic Survey property, Tom looked at his watch and said calmly, "The nuke just went off. The seismic wave is on its way." Oh, fine! *Let's see... It's about five or six hundred miles to Jackass Flats and the compression wave is traveling at three kilometers per second...Hell, I'll figure it out later...*

Tom and I were out of the car and into the cave at a dead run. We flipped switches on at an alarming rate and coaxed the big digital tape recorder to life. The strip chart recorder finally began rolling its paper out with the familiar background wiggles, and then we saw the jerk of the recorder needle as the first nuclear blast shock wave hit. The system had been running maybe fifteen seconds when it arrived. We decided that evening over a beer that we had cut it a bit thin.

We finally finished the successful testing of the digital earthquake seismograph system in September 1963 and sent for the truck. We and it returned to Dallas.

After returning to my home base in the Frito-Lay Building I was involved in an interesting diversion occasioned by a request for a proposal from the government. They were going to be firing a rocket to the moon in a year or two and wanted some interesting instruments to drop to the surface as the rocket looped around it. One of the instruments they really wanted to plant on the moon was a gravity meter that would radio back exactly how strong the gravity was on the surface. This was a job for Boyd Cornelison. Boyd, as I mentioned earlier, was one of the brains behind the Worden gravity meter, now being built by the TI Houston plant. So they called Boyd out to the Frito-Lay building to brainstorm ideas for the proposal. Since Boyd and I were friends, I got invited to the meeting too, as did Fred Romberg, also a gravity expert.

Boyd could easily build a Worden-type gravity meter for the moon out of the fine quartz fibers and springs, but there was a problem. The lunar surface delivery vehicle was to be a large sphere that would be retrofired out of the orbiting rocket and fall more or less straight to the surface of the moon. The gravity meter would be inside this sphere, and it would impact the lunar surface with all the subtlety of a train wreck. I don't recall the expected landing Gs, but it was in the thousands. It was going to hit the moon really fast and really hard. The "shock absorber," if it could be called that, was an outer shell of balsa wood, maybe a foot thick, made to crush on impact and reduce the jolt of the crash landing. This monster jolt posed a problem for the fragile Worden gravity meter and Boyd dismissed it immediately. It would have been smashed into quartz giblets. We needed something that was a lot more rugged. It took Boyd maybe a minute to come up with an idea. "Hey! If we could land an hour-glass on the moon, I'll bet the sand would run slower than it does on earth!" This was the beginning of U.S. Patent No. 3,505,873 titled, "Material Flow Gravity Meter."

Not wanting to waste a minute, Boyd and I drove out to his house on Strait Lane and went first to his kitchen and then to his laboratory. From the kitchen he snitched his wife's three-minute "hour-glass" egg timer, and from the lab, his stopwatch. And I had my official TI-issued Engineering Notebook, into which engineers wrote down important things in ink.

I stood in the middle of his empty double garage and swung the egg timer around my head on the end of a string. Boyd stood by the side holding a protractor in front of his squinty eye getting a reading on the angle of the string as I swung it. A half an hour of egg-timer swinging and stopwatching finally gave us three points on the curve of gravity vs. run time of a three-minute "hour" glass. Sure enough, as Boyd had surmised, the sand ran faster with more gravity and slower with less gravity. Three points on an experimental curve were plenty for anybody but sissies. We were in business.

As junior member of the team of Cornelison, Romberg and Millis, I had to write up the patent disclosure. As usual in my writing, I couched the descriptions in the vernacular, allowing TI's patent attorneys to translate this into patentese. For those of you not familiar with the language of patents, perfectly simple descriptions are turned into obscure and convoluted sentences, totally meaningless to the person doing the inventing. My favorite phrase is a "resiliently biased member" which means it has a spring on it.

My description of the positive features of this gravity meter turned up almost verbatim in the issued patent as, "Although it has several hundred thousand moving parts, when constructed in accordance with the present invention, it is neither fragile nor temperature sensitive." And, strictly speaking, an hourglass *does* have a lot of moving parts, but never try to be funny in a patent disclosure.

119

But whatever the government agency, they didn't like our proposal for a lunar gravity meter and instead, other experiments were chosen to be dropped onto the surface of the moon. Ours was just dropped. Boyd and I parted ways again, but it had been a fun interruption.

And speaking of fun, my good friend Harry Waugh was now in Science Services and we often lunched together and wandered the corridors of the Frito-Lay building during our break periods. And what did we discover as we were wandering one day? The "Wishing Pond" had been closed for repairs, and probably also for scooping out the coins, and was now dry as a bone. It took Harry and I about 15 seconds to think of something useful to do with this odd circumstance. I lay down at the edge and picked up a nice fist-sized smooth round river rock from the hundreds neatly lining the now-dry bottom.

The next day at a magic and novelty shop I purchased a genuine glass-set "gold" engagement ring for $1.49. The "diamond" was at least two carats. Harry's job was to epoxy it to the top of the rock I had filched from the bottom of the Wishing Pond.

A week later the Wishing Pond was refilled and back in business, and Harry and I struck while no one was looking. Using some hook-up wire and a five-foot piece of electrical conduit, I fashioned a snare and cinched up our prize on the far end. Whistling nonchalantly, we approached the pond during the light-traffic morning period, and as Harry stood lookout, I inserted the rock with the ring epoxied on the top of it into the bottom of the two-foot deep Wishing Pond. Continuing to whistle casually, we sauntered back to our offices. Nobody suspected a thing.

Periodically Harry or I would take a quick peek at the pond, but remarkably enough, the "diamond ring" remained undetected for two days. On the third day as Harry and I returned from our usual barbecue lunch, there was a crowd at the Wishing Pond. And one person in this crowd was on his knees poking at the water with a stick. Success at last!

We ambled up and joined the crowd and watched as the would-be fisherman jabbed at the bottom. He had a wooden pole with a nail sticking out of the end and his target looked to be a lady's diamond ring. "It acts like it's stuck to the rock!" he announced to the crowd. Well, I should hope so. That was Harry's job.

And then, up came the pole with the ring dangling from the nail. Not twenty feet away was a jewelry store. "Well," replied the golden retriever, "I'll know in a minute..." Harry and I quietly left.

As 1963 and the digital seismograph contract both wound down to their conclusions, I was shifted over to begin preliminary work on another government project. TI had gotten a contract to modify a LaCoste-Romberg gravity meter so that it could be used on board an aircraft. This would be a great instrument for the military since mapping the earth's gravitational field was a priority at the time. The best instrument they had for gravity mapping was the LaCoste-Romberg shipboard gravity meter, which was, technically speaking, the eighth wonder of the civilized world. But the speed of mapping was limited by the ship's speed. Airplanes were a lot faster, and so the idea was born: why don't you just modify the LaCoste-Romberg unit a little so we can use it in our airplanes? What could go wrong?

I drove to the LaCoste-Romberg plant in Austin twice to learn about their gravity meter system and to develop a plan to make it airborne. It was a very complicated instrument because trying to measure gravity in a moving ship, or a moving anything, was almost impossible. Freshman physics points out that gravity and forces of acceleration from motion are indistinguishable. This is a real bummer when you would like to measure gravity to a part or two per million, which is damn little, and your instrument is jumping up, down and sideways in a ship, causing accelerations thousands of times greater. But LaCoste and Romberg had done it. The instrument, to my eyes, was about 2% gravity meter and 98% corrections to the readings because of accelerations.

Without going into the usual sleep-inducing details, I'll just say that after a month of studying the system, and looking at the contract specifications, I was getting a bad feeling. I had busted my ass on projects in the past, but I had at least entered into them optimistically. Probably too optimistically. And optimism was not the feeling I was getting from this project. It was a feeling of high anxiety. I was badly overmatched. So in January of 1964 I went to Eddie Neitzel with my fears.

121

After hearing my apprehensions, Eddie assured me that it was a viable project and I'd do just fine on it. And he was counting on my engineering expertise to carry it through. This managed to raise my level of anxiety even higher. He was counting on me to make this project fly? I think not.

Across the big room I had been keeping an eye on my old friend Buford Baker, the ex-manufacturing manager of the Houston plant. He had moved up to Science Services shortly after I had, and as usual was in the thick of interesting technical challenges. He was always near the top of my "good guy" list, so I went to talk to him about my quandary.

He had just sold an incredibly audacious proposal to the Air Force Project VELA UNIFORM. It was to build a three-axis seismograph that could be dropped up to 25,000 feet deep in the ocean. In place on the bottom, it would tape record the seismic activity for thirty days. Then, when called like the sleeping family dog, it would shake loose from the bottom and float to the surface for recovery and analysis of the recorded data. This ocean-bottom seismograph could conceivably be dropped off the coast of Russia to record any nuclear test shots that might occur. Great idea! It was an appealing project, even though putting something almost five miles deep in the ocean, running a tape recorder for a month, and then getting it back seemed a little daunting. It was appealing and daunting at the same time. And, oh, yeah, the prototype had to be in the water for testing by summer, like six months away. But it still looked better than the airborne gravity meter project. A *lot* better. I unashamedly begged Buford to save me from a fate worse than gravity and get me on his project. Without any promises he said he's see what he could do.

A week later, Eddie Neitzel left town on a business trip and made the grave tactical error of leaving Buford in charge. The first thing Buford did was to transfer me out of Eddie's branch and into his own. Hallelujah, brother! Buford always did have the direct approach.

By the time Eddie returned from his trip, I was deep in the Ocean Bottom Seismograph project and avoiding eye contact with him. Eddie was a gentleman about it and made no effort to have me assassinated. At least, not that I know of. Instead Eddie assigned Jack Thompson to head up the gravity meter project. He was a much better choice than I was anyway, since Jack was a senior mechanical engineer with years of experience.

The OB Seis project looked like the Three Stooges in Fast Forward. The schedule was so tight everyone was working at a dead run. I had been assigned the sonar recovery system and was hard at work trying to find out how sonars worked. The good news was Tom McCullough was going to perform his digital wizardry on a nasty little signal decoder that I needed.

In the meantime, as I tinkered with breadboards of sonar amplifiers, Buford and others were struggling with the housing for this monster. They had proposed an aluminum sphere about forty inches in diameter as the carrier for the system. Preliminary calculations showed that if it were made strong enough to withstand the ocean-bottom water pressure at 25,000 feet, it would still be light enough to float. Just barely. So they sought out a company that could fabricate aluminum hemispheres about an inch and three-quarters thick and forty inches in diameter.

There was a company in the Midwest that specialized in "spinning" aluminum plates into hemispheres and other strange shapes. Buford and his mechanical engineer went to see them with their plans. The company had no idea if they could do it or not, and particularly on such a tight schedule, but they'd give it a try. The piece we needed was bigger and thicker than anything they'd ever done, and would push their machinery to the very limit, if not beyond.

Several weeks and a number of tries later produced only a large pile of expensive aluminum scrap. But they were getting better. Each piece of scrap was better than the last, but it was still scrap. The hemispheres had to be perfect. The pressure 25,000 feet deep in the ocean is 11,500 pounds per square inch, and perfect hemispheres, heat-treated to the ultimate strength of 6061-T6 aluminum, would have a safety factor of only 1.2. It would just barely be strong enough to keep from collapsing from the pressure. Make it thicker? Then it wouldn't float. The proverbial rock and a hard place.

A final meeting was held in the Midwest company, with the success of the project in the balance. Did they think they could ever produce the big aluminum pieces that would meet specifications? Or should we give up on spinning and try to find some other way to fabricate them? And at the time, nobody had thought of a good second choice. Or even a bad second choice. In theory, spinning the aluminum to form it was by far the best way. It's just that they couldn't do it.

As they gathered for the crucial meeting, the spinning machine operator declined the invitation to attend. He said he was busy. And of course, he secretly wanted to try it just one more time. He had an idea about the temperature and spin speed he thought might work.

So as they met to decide that it couldn't be done, he did it. They walked out of the meeting, discouraged and uncertain as to the next step, and were met by the spinning machine operator. "Come see what I got!" he said. And the project lived.

Somehow, in his spare time, Buford had built an operating model of the crucial 30-day tape recorder at home in his garage shop. The main component that I recall was a Chevrolet engine flywheel. Buford thought some inertia in the tape drive system would help smooth it out, since it was running so slow that you could hardly see it move. The recorder met specifications, and another impossible piece of the project puzzle dropped into place.

As my sonar receiver came together, I began work on the sonar transmitter that was necessary to "call the dog" from the bottom of the ocean. The calculations on a sonar signal traveling through the ocean were not well founded, and varied by temperature and other unknown factors. The best answer I could come up with showed that worst case I needed about 75 watts of acoustic power into the water at the surface to reach five miles down and five miles to the side and call up our big aluminum ball.

The idea was that the "mother ship," hovering in the area of the submerged Ocean Bottom Seis, would send out a coded sonar signal to operate the anchor release mechanism on a specific OB Seis. Then, with the anchor jettisoned, it would float to the surface for recovery. If you had dropped several in a location, you would probably like to call them up selectively, so we used a frequency-shift coded sonar signal that could be preset on each OB Seis. We sure as hell didn't want anybody who accidentally steamed by with their sonar running to pop them up.

So, with the requirement of needing 75-watts of acoustic power now not very firmly established, I set out to build an 800-watt

124

sonar transmitter. A little extra power never hurt. The size was determined by how many transistors I could stuff in it. That was a lot of power for transistors, and during the development I sacrificed quite a few on the altar of the Transistor Gods. I could make the lights dim in the lab when I turned it on.

I finally received the barrel-stave transducer that I had ordered. This was the "loudspeaker" that my 800-watt transmitter was going to drive, and that's what was going to put the sonar signal into the water. It was about ten inches in diameter and five feet long, covered with a thin rubber skin. It could only be tested at full power when it was completely submerged in water, so I took it out to White Rock Lake and rigged for testing.

I brought a borrowed automobile engine hoist and a long extension cord in addition to the sonar receiver, transmitter and the great big black transducer. The White Rock Sailing Club was kind enough to let me use their pier at the north end of the lake for this semi-scientific experiment, and had volunteered their dinghy and small outboard motor.

Using the engine hoist, I dangled the transducer from the end of the pier and submerged it vertically in the lake. I connected it to the transmitter and powered it up. It operated at about 6,000 Hz, which is a really high-pitched whistle. Ear piercing would describe it. I left it running and boarded the dinghy with my little battery-powered sonar receiver and set sail with the outboard motor. Every so often I would stop the motor and drift, and lower the detector over the side and take a reading on the sonar signal. It really wasn't necessary. I could hear the sound of the sonar coming through the bottom of the boat halfway to the other end of the lake. I considered that a success. The sonar transducer did indeed put out an acoustic signal, and a lot of it.

So I added a pointy end on the front and some tail fins on the back of the transducer to make it streamlined and look like it could be pulled through the water behind a ship. A towing frame and a harness of steel cables completed the rig. What could go wrong? I already knew the answer to that one from past experience, so I took it out to Possum Kingdom Lake, just south of Graham, Texas. I rented a cabin cruiser and skipper for the day, and I wore my swimsuit under my clothes.

It didn't take long in Possum Kingdom to determine the towing characteristics of my eyeball design. The lake, a favorite of scuba divers, is clear as a bell. I hung on the side of a rowboat while the cruiser towed my "fish" by. I would duck under water with my dive mask and watch it as it glided through the water four or five feet beneath the surface. Various adjustments of the harness showed that nothing was very critical, so I left the lake feeling that another piece of the project was under control.

Soon, it was the summer of 1964, and as the old saying goes, "Time flies when you're working your butt off." It was time to be off to California for testing. We had made the ridiculous schedule. Now, if the damn thing would only work.

We rented a big ex-US Navy wooden-hulled sub chaser, the *Lois B* and her salty skipper, Captain Huckleberry Cousen. This decrepit old ship could no longer catch subs since its powerful Allison engines, or whatever they were, had long since been replaced with a pair of wheezing diesels, one of which smoked like a Mongolian barbecue. The best feature of the *Lois B* was the wooden deck. It was really easy to screw stuff down to it. The second best feature was Captain Huck himself. He was a real character and a delightful person in his seafaring way, and a genuinely skilled boat driver. We also enlisted the aid of a boat yard at Fish Harbor in San Pedro Bay. It took several days to rig the ship to our liking with an auxiliary generator, a crane, and a really big spool of polypropylene rope. We were soon ready to visit Catalina Island.

The first drop of the Ocean Bottom Seis was about fifty feet off the coast of Catalina in maybe ten feet of water. Better safe than sorry, you know. It was so shallow that the whip antenna of the beacon transmitter was sticking out of the water. But, on command, the sonar signal triggered the anchor release and the big yellow ball majestically rose the remaining few feet to the surface. We greeted it with a mighty cheer. Long journeys start with small steps. Small, safe steps.

We progressed with deeper drops, daily shuttling the *Lois B* between Catalina Island and our dock in Fish Harbor. After a few days of this I was beginning to feel pretty salty and my concerns of possible seasickness began to wane. And then came the day the tide caused a strong current flow through the channel between Catalina and the shore. What had been a smooth trip out turned into the trip from hell on our way back to port in the afternoon. The *Lois B* was bucking into quartering swells several times bigger than anything I'd seen previously. Captain Huck didn't seem concerned and just cautioned everyone to hang on and stay in the cabin.

I sat in the cabin and watched the waves and soon found myself with a mounting feeling of seasickness. I went below and tried not watching the waves for a while. Worse. Up topside in the cabin where I could at least see what was going on was better. It was terrible, but better. Sitting in a chair at the dining table and trying not to think about getting really seasick wasn't working. I was getting really seasick. I could feel the rising nausea. And then the big wave hit.

This wave was big enough to break high over the bow of the *Lois B* and drench the windshield of the pilothouse and flood the deck. She took a hard deep roll to port, accompanied by the crashing of pots and pans that cascaded from the cabinets in the galley. As everyone grabbed for support, overturned chairs sliding across the cabin, Ben Kimler pointed toward the stern and yelled, "The generator!" We all turned to look and saw the 2,000 pound diesel generator skidding sideways on the deck towards the port railing and the deep blue sea. As we watched helplessly, the ship abruptly stopped the roll to port and swung to starboard. The generator stopped in mid-flight just short of the railing, paused for a long second, and then neatly slid back to its original position on the deck.

Captain Huck turned the ship's head into the swells as we raced out on deck to secure the generator. In a few minutes we had tied it six ways from Sunday, something we should have done previously. And remarkably enough, it was back where it had started from except for one thing—one skid was now resting on the big rubber power cable.

We straggled back into the cabin, relieved that our rental generator hadn't taken an expensive and permanent bath, and began setting up the chairs and tables again. Then I noticed. I didn't have the slightest feeling of seasickness. The excitement and adrenaline of the past few minutes had cured it completely.

In the course of several days, we had progressed to a 1,000-foot deep drop in the middle of the channel between Catalina and San Pedro Bay. Ever cautious, a buoyed rope was tied to the cast-iron base anchor of the OB Seis in case of a release failure. We could always winch it back up in an emergency. But, as usual, the unexpected happened and luck was on our side.

After the late-afternoon drop and the deployment of the safety rope buoy, we headed back in to port with the plan of an early morning recovery. And also with the plan of a good dinner and some rest. The beacon receiver we were carrying on board was turned off and secured, since we wouldn't need it until the next morning. Oh, yeah? Captain Huck had turned the ship around and we were headed back when one of the engineers stepped to the stern of the ship to relieve himself. As he idly scanned the ocean during the process, he noticed a big yellow ball surface a mile behind them. *Good grief! It's our Ocean Bottom Seis! Ahoy Skipper! Port your helm or something!*

And so, for an unexplained reason, our baby had shed its anchor and come to the surface. Maybe it didn't want to spend the night by itself on the bottom. Back we steam to the drop zone to recover the errant ball and winch up a 1,000 feet of rope with a big piece of cast iron tied on the end of it. Dinner was late.

We tried the 1,000-foot drop again the next day with satisfactory results, only this time we left the beacon receiver on to warn us if the OB Seis surfaced when we weren't looking. And we all took turns peeing off the stern on the way back to shore.

Now it was time for the grand finale. There was a 7,000-foot deep ocean trench between San Clemente Island and the coast and that was to be our last drop target. That's why we had ten thousand feet of rope spooled on the afterdeck. This was going to be my grandest test also. I proposed that to bring it back up we circle the drop site about a mile to the side to give a slant range to the Seis of about 10,000 feet. I had calculated that I should be able to bring up the OB Seis with seven watts of sonar power at that distance. Seven watts should just barely do it. The slide rule said so, and the slide rule does not lie.

The drop was without incident, although feeding out the rope took a long time. We buoyed it off with a 55-gallon drum equipped with a battery beacon light. It's easy to lose things out in the ocean, and this sure looked like out in the ocean to me. The plan was to circle the buoy all night and let me do the sonar recovery first thing in the morning.

Captain Huck left the ship's engines idling and we took turns all night steering the ship in a big circle, keeping an eye on the faint light that marked our floating drum. Dawn was welcome, as I don't think anyone slept well. And the barrel was right where we had left it.

I made sure the transducer fish was launched and towing properly behind the *Lois B* and powered up the sonar transmitter. The transmitter was set to seven watts, and, as Captain Huck steered the ship in a circle a mile from the barrel, I pushed the "Transmit" button with crossed fingers. The familiar chirping of the sonar signal could barely be heard on deck. I hoped it could be heard 10,000 feet away by the yellow ball. I shut it down after a five-minute sonar call and waited for the OB Seis to surface. It was estimated that twenty minutes would be required for it to float up 7,000 feet. I didn't know if I could wait that long.

Twenty minutes came and went and still no signal on the beacon receiver and no visual sighting. Lord knows, I was looking for it. I had a short conference with Ben Kimler, the project leader. More sonar power? Maybe the slide rule lied. So I upped the power to twenty-five watts, or more than three times the power I had calculated would be sufficient. Bummer. I keyed the transmitter again and the sonar sound on the deck was louder this time. Again the twenty-minute wait. And again, no OB Seis. Well, shit.

I talked to Ben again and told him I didn't know why it hadn't come up, but I had one more thing to try. Full sonar power. *Okay? Okay*, he said. *Do it.*

This time you could hear the sonar chirp plainly on the deck of the *Lois B* as I had the full 800 watts going into the transducer. I gave it a good long transmit and then began anxiously to scan the ocean near the buoy. The twenty minutes seemed like an eternity, but the result was the same. No OB Seis. I went in to the pilothouse and told Captain Huck, "Drive this boat right over the buoy. That's the shortest path between the transmitter and the seis. It's my only hope!" Huck turned to me and said, "Dammit Ed, this is a ship, not a boat!" and then he turned the wheel and the *Lois B* swung towards the buoy.

At that instant, the big yellow ball bobbed to the surface near the barrel and our beacon receiver began to whine. Hot damn! It had *surfaced!* If we'd been closer I would have jumped overboard and kissed it.

We sent the recorder tapes from the 7,000-foot drop back to Dallas to see how the seismometers had worked and they phoned back they had found a problem—the vertical pressure transducer trace had gone away about an hour after it landed on the bottom. We checked it and found that salt water had been forced through the seals by the pressure at 7,000 feet and had shorted out the transducer wiring. That was the best news I'd heard. My sonar receiver signal was tapped off the pressure transducer but, even with it shorted out by salt water, we had been able to drive enough acoustic energy into it by brute force to get the needed 0.2 microvolt. No wonder seven watts didn't work.

The Ocean Bottom Seismograph was a success. Many were built and deployed, and most of them were recovered. And I was greatly pleased that I wasn't called upon to do shipboard duty in what came to be called "The Ball-Bering Sea." Or anywhere else.

I did get to drive a truck from Dallas to Mobile, Alabama, with an OB Seis and accessories to fit up a Coast Guard ship to make a drop in the Gulf of Mexico. That turned out to be a swell holiday cruise for the shipboard TIers as, about the time they were well out in the Gulf, a hurricane blew in and the Coast Guard vessel was diverted into the storm to rescue a fishing boat. That sort of thing is nowhere on my job description.

That particular drop had a strange ending. The seis was not recovered, for whatever reason, but washed up on a beach near Galveston sometime afterwards. The OB Seis was fitted with a magnesium link that would erode in the salt water and detach the anchor after a couple of months if the release mechanism failed. It floated up on a deserted beach and was found by a beachcomber on a Sunday morning. He read the big sticker on the side that said it was a harmless piece of scientific equipment, and that if it were lost, TI would pay $500 to its finder. And it gave a phone number to call. I'm sure he was pleased as punch about this windfall, but must have been faced with a quandary—it was too heavy to put in the back of his pickup by himself, and if he left to get help, someone else might find it. But he was as clever as he was lucky. He peeled the reward sticker off the big yellow ball and took it with him. Now no one else would know what it was or who to call. And they might think it was a naval mine and stay clear of it. It worked. TI got the OB Seis back, somewhat the worse for wear, and the beachcomber got his reward.

Back in the Dallas office I ran into a slightly obnoxious engineer who had been designing an underwater "fish" at the same time I was doing my towed sonar transducer. We had discussed our towing designs and he didn't hide his smirks of disapproval when I told him of my eyeball design and the adjustments in Possum Kingdom Lake. He said he simply designed his on paper, using the proper hydrodynamic analyses. He didn't need to test it.

My slightly obnoxious friend was recently back from a couple of weeks of testing in Lake Washington, near Seattle, so I asked him how his towed fish worked. He replied, "Just fine. And all the data we got from the fish looked good." Well, I thought, I guess he's a good engineer even if he is kind of an asshole.

Another engineer had been on the same trip testing his underwater instrumentation and I saw him a few days later. We began trading sea stories of our recent excursions. Then I asked him about my "friend's" towed fish, and that he'd said it worked fine.

The engineer laughed till his eyes watered. "Worked fine? *Worked fine?* When he first towed it, it jumped clear out of the water like a dolphin and then dove straight down until it nearly broke the towline. He tried a few things that didn't help, and then somebody suggested he tow it backwards. He did and it worked just fine. That's what he was talking about. When he towed it backwards."

My complaining about sea duty didn't do much good. As my part of the OB Seis project faded, I began to ramp up on a project for GSI. It was an interesting problem faced by the GSI marine crews. The marine crews were either one or more often two ships that did reflection seismology in the ocean to find oil. Instead of planting "jugs" in the ground to pick up the reflected vibrations of a dynamite shot as they did on terra firma, the "instrument boat" towed a long plastic tube, called a "streamer," that contained the microphones to pick up the echoes in the water.

A streamer might be several thousand feet long, and could be coiled up on a giant power-driven reel when not in use. When the streamer was deployed behind the boat, the front end of it was held at a depth of twenty feet by a paravane, and then, if things were adjusted just right, the thousand or two feet of streamer following

131

would also be at twenty feet. But, of course, it never was. Particularly in the Gulf of Mexico near the mouth of the Mississippi, the salinity of the water and thus the buoyancy changed enough to either float or sink the streamer from one day to the next. The cure was to reel the whole thing in and add or take off thin pieces of sheet lead along the length. It took time and a lot of black electrical tape.

They really needed a way to automatically adjust the floatation of the streamer and keep it at the right depth all the way from the front end to the back end. If it wasn't right, the signals it picked up were distorted. As you know, there's nothing much worse than distorted seismograph signals, especially if you're trying to find oil.

Being an electrical engineer by degree, or maybe by degrees, meaning a little at a time, I first tried a purely mechanical approach to the "automatic streamer ballasting system," as this problem was called. After messing with several ideas I designed and built a flying wing device. A number of these would be fastened to the streamer at several hundred-foot intervals and they would "fly" it to the right depth.

The device had wingspan of a few feet and was attached to the streamer with a swivel. If the streamer twisted, which it had every right to do, the "flying machine" would stay upright by virtue of a streamlined float above and a weight at the bottom. The angle of the wing was driven by a large sealed stainless steel bellows. As the bellows went deeper the increasing water pressure would begin to mash it shut, and a cunning linkage would convert the movement of the bellows into a change in the angle of the wing. Thus, when the wing and the attached streamer had reached a pre-set depth, the wing would level out and "fly right," so to speak. If it tried to go deeper, the wing would angle to drive it back toward the surface.

So it was off to Lake Dallas with a carload of equipment and the ever-helpful Buford Baker. Buford had a boat in a slip at Lake Dallas and had volunteered to help me do the towing tests. He would have been my choice for help even if he didn't have a boat.

Buford designed and built this boat in his backyard. The construction was brute simple and mostly marine plywood bent over frames. The engine was a converted flat-head Ford straight six. It

132

was about a twenty-two footer with a nice roomy cabin and plenty of space.

Buford ventilated the bilges and cranked the engine up while I loaded the stuff out of my car. I had my automatic ballast model, some PVC pipe to simulate the streamer, a homemade depth indicator and a lot of rope. And, just in case, I had made an emergency float with a long line and a weight to mark a spot if the unthinkable happened.

Eddie Neitzel had showed interest when I mentioned that I had built a remote depth indicator that I could attach to the gadget and then read its depth in the comfort of the towing boat. When I showed him a can of R-12 automotive refrigerant with a can-tapper, a 0-15 psi pressure gauge and a carton of Tygon tubing, he said he wanted to see my depth indicator. I said he just had, except for the duct tape I'd need to install it. He expressed some disappointment in its lack of transistors but I was able to remedy this in a few minutes with the duct tape. I taped three nice new NPN transistors on the side of the Freon can, and renamed the tool, "The Three-transistor Remote Depth Indicator." Eddie was not amused.

I was careful to adjust the buoyancy of my test system so that it would just barely float. I had a fear of it somehow sinking to the bottom of the lake never to be found. Can't be too careful. But of course, if it did sink, I could throw my emergency marker buoy overboard and we could grapple for it later. What could go...? Never mind.

After duct taping everything together, I was ready to cast off and run some towing tests. The depth gauge, regardless of Eddie's disappointment, worked just peachy. The open end of the Tygon tubing was taped onto the "streamer," and then led back up along the towrope to my can of Freon and pressure gauge. The pressure gauge was plumbed to read the gas pressure at the top end of the Tygon tubing. To operate this marvel of low technology, the Freon valve was cracked to give a slow bubbling of gas out of the open end of the tube, which was now under water on the streamer. The gauge would read about a half a pound per square inch for every foot of depth. We achieved a remote-reading 0-30 foot depth gauge for a very small outlay of cash and time, *and* it had three transistors.

But first Buford wanted to give me the nickel tour of his swell boat. Of course I wanted to see the engine, as it idled quietly under the plywood cover. Buford unlatched and raised the cover only to be met with a thumb-sized stream of water coming out of the side of the engine block and going into the bilge. Well, the water pump was working.

Buford turned off the ignition to stop the hemorrhaging and started the electric bilge pump. Inspection of the hole in the side of the engine proved it to be what is commonly called a freeze plug. Or to be more precise, the lack of a freeze plug. Buford fetched his toolbox and looked to see what he had in stock. Two minutes later it was fixed. Buford had reached in his pocket and taken out a quarter. He wrapped a piece of rag around the quarter, doped it good with Permatex, and pounded it into the freeze plug hole with a screwdriver and a crescent wrench. Ready to go towing? You bet, and I'm glad I asked to see the engine first.

We towed the pseudo streamer slowly for probably an hour and were getting reasonable results when, what else, the unthinkable happened. The towed rig hit something solid and came to an abrupt halt in the water, and rather than rip its heart out, I tossed the tow rope and depth gauge over board, followed quickly by the emergency marker buoy. I wasn't a Boy Scout for nothing.

We circled back slowly to the marker buoy for recovery, since I knew the system would float, but in the light chop on the lake, nothing was seen. And then I was horrified to notice the marker buoy floating off in the breeze. The anchor line was too short and the weight hadn't touched bottom and marked the spot after all. Within a couple of minutes, it had drifted away and was useless. I looked at the shore and tried to find some features I would be able to recognize later, like tomorrow. We beat back and forth over the general area until dusk and gave it up. I was fairly pissed, what with the best-laid plans... My precious prototype was going to spend the night, if not eternity, in Lake Dallas. I didn't think I'd mention this to Eddie for a day or two. How embarrassing.

The next day, bright, early and cold, I rented a boat and driver from a local marina since Buford was busy. We headed out to the area that I thought contained my lost equipment, guided by yesterday's landmarks. During the night, a blue norther had brought a cold wind and the chop on the lake was up considerably. Few boats were out fishing in the rough water.

We circled and crisscrossed the area for an hour. I was beginning to get discouraged. I had the skipper cruise the downwind shoreline as close as he could, just in case it had surfaced and drifted ashore during the night. Nothing.

But as we headed back into our search pattern, we spotted the glint of a shiny metal fin in the depth of a wave. We'd found it. If the water hadn't been choppy, it would never have been seen. My machinery was caught in the top of a submerged tree, and it had no opportunity to float or do anything else. Recovery of my autoballast prototype was an anticlimax after the search. I even got my remote three-transistor depth gauge back, transistors and all.

As the year of 1964 wound down, it became obvious that I would have to go to sea again. About the only way to test a seismic streamer gadget is to go where seismic streamers are, and the closest were in the Gulf of Mexico. I arranged for me and my streamer-flying machine to join a GSI crew in action not far off the coast of Texas near Aransas Pass. I packed my bag and my equipment and flew to Port Aransas. I had no idea what to expect.

I was met at the specified pier by the dirtiest boat I had ever seen. This was the "crew boat" that shuttled people and supplies back and forth from the GSI seismic crew in the Gulf. It was already dark when we pushed off and headed out through Aransas Pass to meet the GSI instrument boat.

An hour of wallowing through swells with the smell of greasy food wafting from the galley just about did it for me. I was turning a pale green when the skipper throttled back and we pulled alongside the big GSI instrument boat. With a little coaching from the seafaring types, I left the crew boat and boarded the GSI boat as the swells pitched both around. My suitcase followed a more direct route—an arc through the air onto the steel deck. Welcome aboard.

The GSI marine exploration crews generally consisted of two ships: a large instrument boat and a smaller shooting boat. The instrument boat carried the huge powered reel on the stern to let out and retrieve the half-mile long streamer. An air-conditioned "doghouse" on the deck kept the recording instruments and their operators cool. To me it seemed a big ship, and it had comfortable quarters and good food. It was an opportunity for me to watch offshore seismic operations for the first time.

This was the hardest working bunch of people I had ever seen. The "rules" only allowed "shooting," that is, setting off

explosives for the seismic source, during daylight hours.
Consequently, at the crack of dawn, and it wasn't a very big crack at
that, the first charge was set off with the resultant "Whump!" and
impressive geyser of water. This continued all day, if the streamer
and recording systems worked satisfactorily, until dark. And they
didn't quit until it was plenty dark. Then they stayed up most of the
night working on the equipment. Breakfast was routinely before
dawn. The GSI crews did not let a minute of "daylight," using the
term loosely, go to waste. Parents, don't let your children grow up to
be seismic exploration crewmembers.

Because of the time to get to and from port, they generally
worked ten days straight followed by four days off. Except this
session. The crew had gotten together and arranged to work twenty
days straight and then get eight days off over Christmas. Great idea,
and they were on the downhill side as I arrived on board. Just a few
more days until Christmas and a long holiday with their families.
Except they had just gotten word that, oops, they couldn't do that.
They could only have the usual four days off. Somebody had
screwed up giving them permission. Not surprisingly, while I was on
board, the crew was in a pretty foul humor.

But I was able to get my testing done with the help of the
crew, and during the testing a serious tactical problem became
obvious. Anything more delicate than a large blacksmith's anvil
would suffer severely on the deck of a seismic ship. It was not from
malicious intent, it's just the way things were on shipboard. The
steel stern deck of the instrument boat, pitching in the ocean and
slick from leaking streamers, made putting on and taking off my
"flying machines" an impossible chore. It was a bad idea. Even if
my gadgets escaped damage from the necessary rough handling, it
would slow down the paying out and winding in of the streamer. It
actually flew the streamer pretty well, but that was only half the
story. Rats.

They finished "shooting" their last "line" and reeled in the
streamer and headed back to slip through Aransas Pass and dock at
Port Aransas. And then the pea soup fog rolled in. Not to worry, this
big old ship had the latest in shipboard navigational radar installed,
and as I stood discretely in the corner of the bridge watching the
operations, I could plainly see Aransas Pass ahead on the radar. I
could also see the big blip of our companion shooting boat directly
behind us, which was also faintly visible through gaps in the fog.

And also on the radar was what looked like a bunch of baby chickens following a mother hen. There were four or five other blips on the radar lined up behind the shooting boat. One of the sailors pointed to them and told me, "Those are the shrimpers that don't have radar. They just talked to us on the marine radio and found out we were going in through the pass and lined up to follow in the fog." Good idea.

He continued, "Also, if we're nice to them they're less likely to accidentally cut across our streamer with their shrimp trawls and wreck our equipment. They have the totally mistaken idea that shooting off explosives in the Gulf scares the shrimp down into the mud where they can't catch them. They would be happier if we weren't out here."

I asked the obvious, "*Does* it scare the shrimp down in the mud?"

"Probably," he answered.

Our big steel mother hen made it into Port Aransas safely and docked with the shooting boat alongside. As the crew began their holiday good-byes to each other, the GSI Crew Chief hollered to the cluster of men. "Hey! You can't leave yet! We've got to move ten thousand pounds of Nitramon from the instrument boat to the shooting boat."

Grumble, grumble. They dropped their bags and duffels and turned to do it and get it over with. The Nitramon, the high explosive used in the ocean for the seismic bang, came in fifty-pound cans about two feet long and eight inches in diameter. Let's see, at fifty pounds each that would make about two hundred canisters to be moved. They began moving them with the vigor brought about by being considerably pissed. And with the vigor came a lot of unnecessary banging around. I left in a hurry to find a taxi to the airport.

As they say, back to the drawing board. And this time, I'm the one who said it. But I also got help searching for the next approach. Billy Davis joined me on the project. Billy was an excellent mechanical engineer, and I was to work with his brother Cecil, another fine mechanical engineer, later in the Semiconductor Group.

The new approach was to avoid attachments on the outside of the streamer because of the difficulty in handling. Whatever we did would have to go inside the streamer. This meant that any machinery would have to be crammed inside an already-full plastic tube filled with kerosene. It would certainly be more costly but would pay for itself in time saving.

The soft plastic streamer tubing was full of plastic spacers, steel strain cables, multi-conductor electrical cables to bring the signals back to the ship, pressure seismometers and kerosene. The streamer was cleverly designed to be filled with a fluid appreciably lighter than seawater, namely kerosene. And even with all the stuff inside the plastic jacket, it would float, if just barely. A few strategically placed strips of thin lead would make it exactly neutrally buoyant. At least for a little while, until you hit the mix of fresh water from the Mississippi, or the Gulf warmed up a few degrees and changed the density of the seawater. But we found enough space inside to put some machinery.

Our plan was to put a tiny electric motor-driven pump inside the streamer with an attached reservoir of kerosene. Running the pump one direction would take kerosene out of the reservoir and pump it into the streamer jacket, which was already full of kerosene. The jacket would expand, and the streamer would get lighter in the water and float a little higher. To make it sink, we would reverse the pump and suck kerosene out of the jacket housing and pump it back into our reservoir. A pretty good idea. The streamers already had remote-reading depth gauges every so often along the length, so an operator could then trim the adjustable sections manually, by running the pump motor one way or the other, until the streamer was at the right depth.

There were two difficult problems. The first was stuffing this tiny motor with its pump and reservoir into the streamer. And of course, having to run submerged in kerosene added an additional challenge to the electric motor. The next and equally serious problem was getting sufficient electrical power to the motor. The streamer had to be kept light to float, and one way to keep it light was to use tiny wires for the electrical signals. Copper is pretty heavy, and a bundle of wires to bring the signals from a lot of pressure microphones gets big. So the wiring in a streamer is tiny wire which goes through a connector at every streamer junction.

Running an energetic pump motor through it would be like trying to jump-start a Mack truck with a pair of clip leads.

One of Billy's early experimental models supplied some entertainment, however. He wanted to try using compressed gas to power a variable flotation device. If the gadget sank too deep, it would valve a little gas into a bladder and make it float more; if it floated too high, it would release gas. An R-12 refrigerant can powered his model, probably the same one I used in my depth gauge. The flotation bladder was made from a bicycle inner tube, and a depth-sensitive valve controlled the flow of gas.

The experimental model was dropped into the fifteen-foot deep sonar test tank at the Lemmon Avenue plant. It sank straight to the bottom and lay there like a stone. Billy and I peered through the murky water and tried to see what was going on. Billy finally said, "I think it's moving. I'm pretty sure the bladder's inflating. Yeah! Look! It's rising off the bottom!"

And rise it did, gaining speed as it neared the surface, the inner tube swelling ominously. It broke the surface and fell back with a splash, floating high on the water. By this time the depth control valve had opened to release the pressure and the assemblage began to make flatulent noises while writhing around on the surface. Finally, with a last feeble *blat* and a flaccid inner tube, it again sank to the bottom of the tank.

We watched the cycles of pneumatic life and death throes, fascinated, until it ran out of Freon and stayed on the bottom. Just like a stone.

In February of 1965, I was scheduled for another information-gathering cruise with a GSI Marine Seismic Crew in the Gulf of Mexico. Buford Baker went with me on this trip and we were to meet the crew boat in Venice, Louisiana. Venice was (since it later blew away in a hurricane) as far as you could go down into the Mississippi delta. The road, such as it was, ended in Venice. It was a fascinating automobile ride through the bayous of the delta region.

The crewboat ride out to the GSI ship took us by dozens of oil drilling and production platforms. We boarded what I recall as the M/V *Floodtide*, which was a big, modern ship. The accommodations were good, the food excellent except for the hour of breakfast, and the crew pleasant and obliging. In addition to the ship's crew and the GSI crew, a hired hand from an English navigation company was also on board. He, with his magic box, could tell the position of the ship within a few hundred yards. His company had the only viable method of doing that in 1965, so he and his company were paid a lot of money. He would be given the day's shooting line and firing schedule and he would then direct the skipper and the GSI man with the firing button where to drive the boat and when to set off the charges. Today the same thing can be done an order of magnitude better with a hand-held Global Positioning System for $89.95.

This instrument boat was big and it carried a lot of powder on its deck. The mouth of the Mississippi was a hard bottom area and large shots of explosives were required to get through it and get a reflection back to the seismic recorders. They were shooting four cans of Nitramon at a time, a 200-pound charge, and they shot them every couple of minutes as we steamed along. In the course of a week, they shot a lot of "powder," as the crew called it. When Buford and I boarded the *Floodtide*, there was about 50,000 pounds of Nitramon stacked on the deck like cordwood. Periodically they would stop both vessels and transfer a load to the shooting boat.

This time the shooting boat cruised along even with the stern of the instrument boat and a hundred feet off the starboard side. The shot and its firing line were dropped overboard from the shooting boat and then fired when it was not much more than a hundred feet behind it. Two hundred pounds of Nitromon would make quite a bang and send a geyser of water probably two hundred feet in the air. It was very difficult to take a nap while this was going on.

Buford and I successfully completed our all-expense paid cruise of the Gulf of Mexico, Mississippi Delta Region, and returned to our own little beds in Dallas. I was slowly but surely beginning to catch on. Everything I worked on in Science Services ended up in the ocean. I really needed to think about this. I didn't particularly like traveling, much less ocean travel. Could it be that my cycle of engineering wanderlust was starting again?

Jim Nygaard had begun his latest attack to get me to come back to work for him. I first got a call from him saying he'd left a message at my house. Knowing that I had built a metal detector a few years before, he thought he'd make it a little more exciting than

140

say, putting a note in my mailbox. Jim had put the missive in a metal capsule now buried somewhere in my front lawn. He knew I couldn't resist, and of course I couldn't.

It took me three evenings of searching and digging to find it. I located and dug up nails, bolts, pieces of roof flashing, parts of toys, a penny, a quarter, and finally, Jim's finger-sized steel capsule. I immediately took it back to my shop.

The cap on the end of the tube had wrench flats on it, so I clamped it in my vise and got my old family heirloom crescent wrench to unscrew it. Let me just say that the cap and flats were a decoy and the end I tried to unscrew was a solid piece of steel. Nygaard still owes me a new crescent wrench. In a sudden fit of pique I sawed it and most of the paper message in half with a hacksaw. The message said something like, "Ed, come back. We love you! JLN" I thought he had a funny way of showing it.

I finally made up my mind. I told Mary Ruth I was going to leave of Science Services and go back to work with good old Jim Nygaard. She rolled her eyes and said, "Not again!" But she loved Jim, too. She just had a funny way of showing it.

Chapter Seven

Jim Nygaard's transistor test equipment group was now in the North Building, having moved out of the Semiconductor Building while I was away playing in the water at Science Services. The North Building had been built because the SC building was absolutely stuffed full of transistor-making machinery and people, and the integrated circuit manufacturing was also beginning to swell up. So Mark Shepherd and all the corporate functions, and Jim Nygaard's group, and a bunch of others moved out and into the new North Building. The fact that I came back and started to work on April Fool's Day, 1965,was merely a coincidence.

In the three short years I had been out of the semiconductor test equipment business, things had changed. The integrated circuits, commonly known as ICs at the time and computer chips now, were catching on big time and production ramping up. Testing needed to ramp up too, and what we'd learned in transistor testing wasn't always applicable to integrated circuits. Transistors had three wires to hook up and talk to, and integrated circuits had more. A *lot* more.

So by 1965, TI needed a better way to test integrated circuits. The problem was not so much technical as tactical. The electrical tests were usually not any more difficult than those performed on transistors, but there were so many of them. The complexity of the internal workings of the IC led to a ton of variations and a lot of testing. In addition, the number of connection pins on an IC package ranged from eight to several dozen, and the function of any given pin, with a couple of exceptions, could be anything depending on the whim of the IC designer.

In late spring of 1965, Lee Blanton, Bob Stadtler and I began the design of a new integrated circuit electrical tester. There was a pressing need for a tabletop tester for all types of integrated circuits, so we designed a compact paper-tape-driven IC tester. It was officially known as the MSB-1100. The "MSB" stood for "Millis, Stadtler, and Blanton" and the "1100" nomenclature was actually taken from the time we liked to slip off for lunch. Plus it sounded kind of scientific.

Lee and Bob, as usual, did the detailed circuit design for it with, as usual, splendid results. Bob Clunn, brand-new Rice graduate of June 1965, also hired in and signed on to help. This would be his first "real job" at TI, having worked two summers previously as a summer development student. I helped by wringing my hands and getting my old childhood friend and also Rice grad, Neal Lacey, to do the industrial styling on the cabinet and controls. When Neal, an architect by trade, got through it looked really nice. Engineers may not have enough *artiste* in their souls to design pretty things but we know people who do.

A key element in the MSB-1100 was the paper tape reader. Mounted at the top of the three-foot tall cabinet, this tape reader was loaded with a continuous loop of special inch-wide Mylar nine-track punched tape. This tape loop had all the test parameters and pass/fail information coded into it that the tester needed to test one type of IC. It would run one loop and stop each time a unit was tested. In production, the test person running the MSB-1100 would receive a box of integrated circuits and the roll of punched tape together. They would simply load the tape in the reader and test the units. It was a nice system.

The tape reader model was chosen with some misgiving because it was the only one in the world we could find that would read tape at the rate of a thousand characters per second. Our fears were groundless and the reader worked well. Believe me, a thousand characters a second is really whizzing the tape through the reader. The more common "paper" paper tape would shred after a few hundred loops. The Mylar "paper" tape would run thousands of loops without destruction. It was also very popular at Christmas time for wrapping gifts since it was gold on one side and green on the other. Very decorative.

The first MSB-1100 was installed on the integrated circuit test line and officially put into production. Before we could walk back to our desks, it had crapped out. We fixed it, and it crapped out again. This scenario was repeated. The MSB-1100 had a serious problem with the plug-in reed relay packages, like they didn't make reliable contact. We found out during the autopsy that the contact pins had not been properly heat-treated. The "MSB" now stood for "Miserable Son of a Bitch."

But the relay contact problem was soon fixed and the tester went on to become a mainstay in IC testing. It worked so well that TI decided to make an outside product of it, and it was renamed the

"TI-668 Integrated Circuit Tester" for its public persona. It was kind of neat seeing a real sales brochure for something you'd worked on.

No sooner was the MSB-1100 under control than we began trying to unravel the next integrated circuit testing problem. This problem was testing the silicon wafers in the "probe" area. The "probe" is where completed silicon wafers with hundred of identical devices are "probed," meaning connected up electrically with tiny needles, and electrical tests performed to determine if the individual devices are good or not. If the device tests bad. a dot of ink is put on it and later, when the wafers are sawed or broken apart, the inked ones are trashed. But it had the same problem as the MSB-1100. It needed to deal with a lot of leads on the devices and perform a lot of tests. Just to make it more difficult. we agreed that the new machine should be able to service several "probe machines" at the same time to be cost-effective.

Probe test systems at the time were "hard-wired" to perform certain tests on certain pins of the integrated circuits, but standardization of the pin functions wasn't possible. Each new type of integrated circuit needed either a new tester built for it, which wasn't feasible. or an adapter made to get the wires hooked up right. Pretty soon, you're spending a lot of time making adapters.

In a weak moment it occurred to someone that if we could design a magic box that could be connected to a pin on an IC and then by digital control be turned into any kind of electrical "thing" we needed, we wouldn't have to build adapters. If we wanted to test a fourteen-pin IC. we would connect fourteen of these magic boxes to it. one to a pin, and write a little software control program. And, as if by magic (we're always needing magic to help our testing), the IC would be hooked up with power on the proper pins, input signals on its inputs. and output loads and measuring gadgets on its outputs. No muss, no fuss. Simple but effective it seemed.

We sat around in meetings for several weeks and kicked the idea around. This magic box had to look like a supply voltage of either polarity, a ground, an input logic signal, a logic output load with measuring capability, or a resistor and a bunch of things I've forgotten about. You then hooked up one of these identical universal boxes to each lead of the device to be tested and electrically told it what to be or do. It became known as the "tester per lead" concept and it was a dandy. Then all we had to do was figure out how to multiplex this around so it could service a bunch of probers simultaneously.

In one of my last meetings on this test system, we were mulling over some problems of the system design, when I asked the group what we should call this thing. One of the attendees said, "I don't know what we ought to call it, but it's going to be one high-speed mutha!" In less time than it took the laughter to die out, the name of HSM, for High-Speed Mutha stuck forever. The HSM Project was now officially under way.

I was chosen to present this exciting (to me, anyway) project in a planning meeting in the Executive Conference Room in an attempt to spread my infectious excitement to the people with money. After my "foil" (everyone else in the world calls them "transparencies") presentation, I left the platform and was passing by Department Manager Jim Reese in the semidarkness on the way back to my seat when Jim caught me by the sleeve and motioned for me to bend down. He whispered in my ear. "What does HSM stand for?" I whispered back, "High Speed Mutha."

"HAW! HAW! HAW! HAW!" Normally mild-mannered Jim cracked up and brought the meeting to a halt, but wouldn't tell them what he "thought was so damn funny." Later, in polite company, it was allowed that HSM stood for "High Speed Machine." But I know differently.

This meeting was also memorable in another way. Sometime during the hours-long conference, Claude Head got up and slipped out to the restroom. When he returned, Cecil Green, elder statesman and revered founder of TI, was sitting in Claude's chair. Claude bent over and said in a loud whisper, "That's my seat you're sitting in!" Every horrified eye in the room turned to watch the outcome of this sacrilege.

Mister Green looked up, surprised, and said, "Oh! Excuse me!" and got up and moved to another chair. It was fun to tell Claude later what he'd done. Me and about ten others.

The design of the HSM was begun without a commitment from the integrated circuit department. What? Us worry? The system was to encompass any IC with up to 32 leads, which was huge at the time, and with the capability of servicing eight wafer probers simultaneously. The test throughput would be tremendous, far above anything contemplated at the time, and the IC management would snap it up as the bargain of the year. Well, not exactly. The resident Industrial Engineer in the Integrated Circuit group ran his own numbers on cost versus speed and said flatly that it wouldn't pay out. Our numbers showed that it would, and in spades, but since he was the "expert," he won. The IC department turned their noses up and their thumbs down on the HSM.

Big oops. The HSM design was well along and the customer just axed our product. But, fear not, for the transistor group had been looking over our shoulders and liked what they saw, and they didn't have an Industrial Engineer. So, quicker than you can say Lee Blanton, the design was modified from 32 leads and 8 probers to 8 leads and 32 probers. Voila! The HSM I was born! A transistor probe test system that would go like blazes and run a whole room full of probers. Blanton always could do the magic we needed.

Lou Broussard, one of the original team members during the concept phase, drew the lucky bean and was project engineer on the HSM I. He was creative to a fault, and got it into successful production and started the test system on the road to a long and fruitful life at TI.

Broussard also made a swell victim. A few years later, our family had a piece of property in East Texas that occupied a lot of my weekend time, what with trying to drill my own water well, which required dynamite, and other interesting projects. Lou also bought a piece of rural property that was very nearly on my route to East Texas, so without great difficulty I could drop by Lou's place and see what he was doing. He had already attracted the attention of his rural neighbors by trying to burn off a patch of weeds and nearly destroying the entire county. He met a lot of new friends that day.

To help Lou meet even more friends, I lettered a sign on one of my wife's discarded oil-painting canvases and nailed it to Lou's gate post early one Sunday morning as I drove out to our place. The sign said simply, "FUTURE HOME OF THE VAN ZANDT COUNTY NUDIST COLONY."

Some days you have all the luck and that was one of those days for me. Apparently minutes after I'd nailed the sign up, a widow lady that lived down the lane drove by it on her way to Sunday School.

During church, she stood up and announced to the horrified congregation what that city fellow was getting ready to do to their community. Community leaders promptly organized and called the realtor who had sold the property to Lou. It was the realtor's damn fault this was going to happen. The realtor, whose livelihood depended on being a friend to all in Van Zandt County, called Lou at home Sunday afternoon. After some small talk to break the ice, he asked Lou about the nudist colony he was starting up. Lou didn't have to fake his ignorance, and it soon became obvious that he'd been had.

Lou called me Monday with strangely psychic accusations and the jig was up. But I like to think the widow lady had some excitement in her life and was maybe secretly disappointed in the outcome.

The transistor department's HSM I probe system ran like a house afire, and soon the IC department regretted their hasty decision not to sponsor it and came back to talk. What they really wanted, they said, was an HSM just like the one the transistor group had, only cheaper. Negotiations continued until the dollars vs. functions were settled. Instead of a pure "tester per lead' that the HSM I had, they wanted a simplified cross-point matrix switching system that would be cheaper and nearly as fast. As usual, the Golden Rule applied, like who ever has the Gold makes the Rules. Thus, the HSM II came into being.

The HSM II still used the "tester per lead" concept on a smaller scale, but it was decided to upgrade the functionality of this magic box. Lee Blanton began to design a new one.

Lee, our chief operational amplifier and other things circuit designer, struggled with hypothetical designs of this new and improved magic box for several weeks, almost cracking the code but not quite. It was a very difficult design problem, and if Lee couldn't do it, I was at a loss where to go next with the problem. I personally couldn't op-amp design my way out of a paper bag, so I needed to find someone else who could design op-amp circuits on a level with Blanton. A daunting and unlikely task, to say the least. At least it was until I spied Tom McCullough wandering through our area. Tom, as you recall, was a TI technical over-achiever and friend from my Science Services days. I immediately cornered him and presented with the sticky wicket. Tom, ever the engineer, promptly

147

dropped whatever his boss had told him to do and took up the challenge.

The next Monday Tom dropped by my desk and on the back on an envelope, honestly, he had sketched a complicated op-amp-loaded diagram that would do the trick. Blanton acknowledged that Tom had indeed cracked the code. Coming in on a tough problem with a fresh and uncluttered brain can be an advantage. A crucial piece of the design had fallen into place, albeit after a hard push. The HSM II development and the construction of production models for the integrated circuit department progressed on schedule.

Later, with minor modifications, the HSM II became the HSM III Eventually over 400 systems were built. As of this writing in the year 2000, many of them are still running in TI plants around the world. It was a durable tester. I was only involved in the first one percent of the design, but that was the fun part because I got to help name it.

Speaking of durable testers, the HSM I in the transistor department, the only one that was ever built, had been fitted with a "unit counter" to keep track of the total number of transistors tested from the day it was installed. An electronic circuit counted to a thousand, then sent one count out to a Veeder-Root mechanical counter. The Veeder-Root counter finally broke, so how many millions and millions of transistors were tested on this machine is unknown. But I can safely state that it takes a lot of counts to wear out a Veeder-Root.

Regressing a bit, during the development of the MSB-1100, I was honored for having served TI faithfully for a period of fifteen years. In August of 1965, I had my fifteenth anniversary celebration, as devised by my peers and not the management of TI. One of the wonderful gifts I received was a "bull wheel." It was to become a memorable and lasting gift.

Years ago, I had picked up the phrase "Johnson rods and bull wheels" from Harry Waugh. I used the words to describe anything mechanical that I didn't understand, and it had become associated with me. They thought it fitting to give me a bull wheel of my very own. It actually was not a bull wheel. A bull wheel is the main gear in a drive system, and this gift I received was not a gear. It was the rotating counterweight from a punch press. At least that's what they were told at the scrap yard where they bought it. It was cast iron and about eighteen inches in diameter and three or four inches thick. A six-inch hole ran through it, offset from the center by several inches.

It must have worked well as a counterbalance, being as lopsided as it was. My first difficulty came trying to get the "present" back to my desk. It could not be tipped up on edge and rolled—it was so unbalanced as to be impossible and not a little dangerous. It weighed in the order of 200 pounds and it was a superior finger-masher. But I persevered, and the bull wheel sat on the floor alongside my desk for months. That is, it did until Jim Nygaard began to bug me about it.

Since he's the one who had willed it into being in the first place, I thought it was a little out of order for him to complain about it rusting quietly on the floor in the engineering area. It wasn't bothering anybody, and the sweepers and moppers just worked around it when they cleaned our lab. But Jim insisted on two things: first, I was to get it out of the building and to my house by the end of the week, and second, no one was to help me move it, under penalty of terrible things. Well, wasn't *that* just great.

Saturday, after I'd put in my morning work time, I snitched a two-wheeled dolly from the facilities people and loaded my bull wheel on it. The wheel left a matching rust image on the vinyl floor from being mopped-around for months. Down the elevator and out the door caused no serious muscle strain, but when I got it to my Mighty Morris Minor 1000 in the parking lot, I had another problem. I had to clean and jerk 200 pounds into the trunk. Or maybe it was a bench press. In any case, it took every ounce of strength I had to drag it up and across the back bumper and into the Morris boot, since English cars don't really have trunks. And I might add all of this took place with several of my "friends" standing by and watching.

The Morris drove funny on the way home, like the front wheels were barely touching the ground, but I reached my driveway and backed into it. A 2 by 12 plank served as a ramp to skid the great iron lump down to the concrete where the trip ended. Thank heavens.

The following Monday I entered the North Building on the usual path to my desk and was intercepted by one of my coworkers.

Breathlessly he informed me that Nygaard wanted to see me, *right then*, and to go straight to his office. *Good grief! What now?*

I'll confess I don't remember the details, but good old Jim kept me in his office and on the defensive for twenty minutes. At any given time in my career, Jim could find enough dubious items to merit an ass-eating, and twenty minutes wasn't too bad. But I really thought I'd been unfairly attacked, if that's the word. So, slightly puzzled, I headed back to my desk to begin my day's work.

As I neared my desk, someone asked me what Nygaard had wanted so urgently, and I said, "Funny you should ask...," and I proceeded to describe my ass-eating with the idea of getting another and more favorable opinion of my work. As I pled my case, I looked down and noticed that I had automatically rested my foot on something by my desk. I had one foot on the bull wheel. The bull wheel was in the exact location I had taken it from Saturday morning, perfectly aligned with the rusty outline on the floor. It was as if it had never left.

It caused a sudden shock to my brain. Did I take it home Saturday or did I just dream it? Do I need professional help or have I just been working too hard? A roar of laughter interrupted my private thoughts. "Millis, you should have *seen* your face when you saw the bull wheel!"

Mary Ruth told me the whole sordid story when I got home that night. As I drove out of our driveway that morning and around the corner, four men, previously hidden in the bushes, sprung forth and ran for our garage, scaring the hell out of Mary Ruth. It was a team led by Lee Blanton, coming to retrieve the bull wheel. After identifying themselves, and then of course with Mary Ruth's blessing, they loaded the bull wheel into their car and headed back to TI. You'll note that it took *four* of them. And of course, the rest is obvious. While I was getting my pseudo-ass-eating from Nygaard, they re-installed the bull wheel on its rusty shadow. Geez.

The details of the next bull wheel caper have faded from my memory, but somehow it was "stolen" from me and returned as a potted-plant holder. Wheels had been added to the bottom, along with a chain to pull it with. It had a nice coat of black paint with gold accents. My office *Dieffenbachia* pot just fit in the shaft hole. It was really a pretty nice plant holder except for one thing—they had boogered up the casters so it wouldn't pull straight. They just had to mess it up. But it became the holder of my *Dieffenbachia*, common name, dumb cane. How appropriate.

To follow the bull wheel to its conclusion, the world's heaviest potted plant holder was moved, along with the plant and my desk, chair and files, to the brand new South Building. Jim's entire Branch moved into nice new quarters on the second floor, south side. I was moved even farther since the space wasn't quite big enough. I ended up "temporarily" in an architectural setback off the upper floor outside aisle. What a great place. No telephone wiring, tons of foot traffic, no one in my group within shouting distance, and birds. Several sparrows that were trapped inside the building liked my space too, and spent most of their free time perched above me in the girders. The good news was that I was leaving on vacation. Before I left, I wrote a "funny" request to facilities complaining about the birds, and requesting them to construct an official bird perch and feeder.

When I returned from vacation I found that my 200-pound cast-iron *Dieffenbachia* holder had been converted into a bird perch. A six-foot length of ten-inch steel pipe had been welded to the top of the bull wheel. On the pipe, near the top, was screwed a tiny rod and birdseed holder, sufficient for one bird to perch on. It had been upgraded from a 200-pound plant holder to a 300-pound bird perch. Lesson: never write a "funny" memo and leave town.

The bird perch was too big for my office. One of us had to go. I made several attempts to get rid of it, but to no avail. Taking it down to the shipping dock and abandoning it did nothing. Nygaard's cohorts kept an eagle eye out, and promptly retrieved it and reinstalled it in my office. But then I had an idea. The South Building was unique in that it had a basement that was full of nooks and crannies. I thought I could probably find a nook, if not a cranny, big enough to hide a bird perch. I began searching stealthily at night, undetected by the Nygaard Mafia. I soon found what I was looking for. Under a stairwell and behind a door, I found a collection of huge TI outdoor Christmas decorations. Perfect. In the dead of night and in utter secrecy I trucked and elevatored my bird perch down to the basement and parked it outside the chosen location. I moved the decorations from the big closet, rolled the bird perch to the very back, and restuffed the decorations. The bird perch, for all practical purposes, vanished into thin air.

For days following, I answered questions about the whereabouts of my bird perch with only a smug smirk and silence. I understand several posses were sent searching for it, but no luck. At last, peace and space had returned to my office.

For a couple of years, I told no one, especially Mary Ruth, and it continued to be a subject of speculation. And then I told just one guy. One lousy guy. He was an outside salesman with whom I was slightly acquainted, and he'd heard the story and asked me, off-handedly, what I'd done with it. And I, off-handedly, told him I'd hidden it in the basement. What could it hurt?

Well, if I'd known that this salesman was the friend of an old high school buddy of Jim Nygaard's, and that the three of them would end up in a fishing camp on the Little River near Honobia, Oklahoma, a few weeks later, I would have kept my mouth shut. After a few drinks and fishing stories Saturday night, Nygaard proceeded to tell them the "bull wheel story," and ended it with, "And then Millis hid it somewhere and I can't find it."

The salesman made Jim's whole weekend with, "Well, I know where it is!" And of course, he did, damn it.

Jim put together an Emergency Basement Bird Perch Search Team Monday morning, and before long the bird perch was triumphantly returned to my office. I lost yet again, me and my blabbermouth.

The end of this long-winded story was an anticlimax. A month later TI was having a plant-wide cleanup drive. One evening as I left work, I rolled the bird perch into the aisle and marked it "Trash." The next morning it was gone forever, never to be seen again. But nobody believed me when I told that part of the story. They still think it's hidden in the South Building somewhere.

The focus of my new job in the semiconductor test business had obviously swung from transistors to integrated circuits. The Super CATs were doing their job and except for several major upgrades over the years, lasted as long as transistors did at TI. My next challenge was to design something to automatically "handle" the "Mech-Pak" device. A "handler" was the gadget that automatically loaded a device that was to be tested, made electrical contact with its leads, and then put the device in the proper category as instructed by the test electronics. It automated the "test and sort" routine.

The "Mech-Pak" itself was a little open-centered square plastic frame about the size of a postage stamp. It was designed to hold a "Flat Pack" integrated circuit, which was a tiny glass and metal package about an eighth by a quarter of an inch on a side, and very thin. Gossamer gold-plated leads radiated from its edges, so delicate as to be impossible to plug into a test socket. The Mech-Pak solved the problem by tightly wrapping the flimsy leads around its outside edges, which hung the integrated circuit in its center like a heretic stretched on a rack.

At the time I was requested to look at automating the testing, the handling was strictly manual. A test person picked up each Mech-Pak, pressed it into a special socket, and energized the tester. After the tests were completed, the operator pulled it out of the socket and put it in the proper bin as indicated by the tester. Just like in the early days of transistors. Lots of opportunities for screw-ups, and for what the ICs were selling for, it behooved us to get them in the proper buckets.

The Mech-Pak integrated circuits, which were high-dollar and mostly for military uses, sorely needed a test handler. To add an additional degree of difficulty to it, it needed to be able to do the testing at a high temperature. The part of this project that I remember best was coming home and sitting in my easy chair with the evening newspaper. And then folding up the newspaper and drawing sketches on it of a Mech-Pak gating mechanism that I'd been puzzling over. *Why sure! That's the way to do it!* I sometimes think my brain works best when I leave it alone.

A prototype of the Mech-Pak Handler worked pretty well, and after a few changes, several of them were built. The first user in the integrated circuit final test area liked it a lot. At least, he liked to watch it run. He would load a magazine of Mech-Paks into the slide on the top and turn it on. The units would be pulled into the test socket one at a time for testing, and then dropped out the bottom of the machine into "good" or "reject" buckets. What my customer liked best was the way the units would be "swallowed" up into the socket. He said it just looked like someone eating Mech-Paks, and he stuck the name "The Gobbler" on my creation. He was right; it did look like it was gobbling Mech-Paks.

153

About this time, the DIP, or Dual Inline Package, for integrated circuits became the rage. It was cheap to produce, easy to handle and convenient to solder into printed circuit boards, which was the exact opposite of the Flat Pack IC. The DIP units also needed a handler, and since I'd just gotten The Gobbler running, I resharpened my pencil and made a few changes. Now it was a DIP handler, much like the Mech-Pak handler but lacking the high temperature capability. DIPs didn't need no lousy temperature testing—they were the rough-and-ready industrial devices.

This new DIP handler soon propagated through TI and to a lot of overseas sites as "The Millis Handler." Years later, I would meet people from the overseas plants, and when they'd see my badge, would ask, "Did you make the Millis Handler?" I thought that was nice.

Shortly after the Millis Handler came the Stadtler Handler. Bob Stadtler designed and built a system for a tough test job that required cooling the integrated circuits to well below zero while testing them.

Bob built a well-insulated test chamber and handler and chilled the inside with CO_2. Anyone who has ever cooled a wastebasket of beer cans with a CO_2 fire extinguisher knows how effective that is. Bob took the latest high-tech approach for the brains to control the actions of the machine and custom designed a digital logic control system for it. This included minimization techniques and all the neat stuff learned in school. His exercise with hard-wired logic for machine control was to give us a good lesson on how to make machines run.

Bob did a textbook job on the logic control system, and it was awful. No one had been able to predict accurately ahead of time what controls and functions and timing would be required to make the system run properly in production. The problem of hard-wired logic, especially minimized, was that any change in function, like adding another control switch or button, screwed it up totally. You had to start over if you changed any part of the system. And guess what—we changed things a lot when we built the first model of anything.

But Bob got it running after much patching and rewiring on his logic, and installed the system in the Final Test department just

as the Christmas holidays were beginning. He had finished training
the operators and the machine was running up a storm as Bob
prepared to bid us adieu until next year. Bob turned to leave, then
went back to his new baby, opened the rear cabinet door, and began
to write something on the panel. I could see, and it said, "In case of
problems, call 555-8383 anytime day or night."

   I told Bob I thought that really showed his sense of
responsibility for his new machine, and I was proud of him for it. He
turned to me, puzzled, and said, "For putting Dial-A-Prayer's
number in the back?"

   The Stadtler Handler was just the last straw in trying to use
"paper designed" logic controllers in test systems. Since we didn't
know exactly what the machine controls would be until after the
machine was built, it was obvious we needed something a lot more
flexible. A *lot* more flexible, like maybe a computer. The price and
size had been coming steadily downward in the past few years with
the advance of integrated circuits, so maybe it was time to take a
closer look at them for machine control.

   Fortunately, Jim Nygaard in all his wisdom had decided the
year before that we needed to learn more about computers. Dr.
Warren Rice was once again tapped to help us out of the unknown.
He just happened to have a friend who taught Computer Science at
Arizona State University, and he had a feeling he could get him to
teach a bunch of TIers all about computers.

   In the fall of 1967, several groups of TIers, with Nygaard
leading the pack, went for two weeks of computer training at ASU
under the incredible talent and tutelage of Brian Thomson. It was
just what we needed. We learned to program in Assembly Language,
the down-and-dirty language of the heart of the machine, and also in
BASIC. It was a great course, with the two-week grind pleasantly
interrupted by a two-day automobile trip into the Four-Corners
country, which was as unforgettable as the computer training. It
accomplished exactly what we desired. It sure looked like a
computer should be able to run a machine, although no one to our
knowledge had used one for that purpose.

   That's when Bill Rylander, one of Nygaard's ace EEs,
jumped in and built the Model Ten computer. In the spring of 1968,
Bill had it running. It fit in a tabletop cabinet along with 4096 words

of 16-bit memory. It used a subset of the IBM 1130 instructions and could do all manner of neat things. Bill said he originally called it the "Model T" computer, because of its minimal instruction set, but Nygaard objected so Bill added the "en" and fixed the name.

Bill was, and is, one of the guys that can do difficult things with apparent ease. How did he know how to build a computer? He just did it.

The Model Ten could be run by hand loading your machine-language program in through "bit switches" on the front panel. That, my friends, is as fundamental as you can get, but it sure is fun. After some brief instructions from Rylander on how to write the code for it, and how to get the code into the memory, most everybody in our group left on vacation, being the summer of 1968. I stayed, and with no pressing projects, I decided to learn how use the Model Ten computer to run stuff. It was one of the more entertaining two weeks I ever spent at Texas Instruments.

Programming it was easy. First the flowchart of the process steps was generated, showing all the loops and forks that the "process" could take. This took the most time, since that was basically the design of the program. After the flowchart was completed and polished up, a string of instructions was written in Assembly Language. The lines of Assembly Language were nearly one-for-one with the symbols on the flowchart. Then, the Assembly Language notations were converted into hexadecimal (base sixteen arithmetic) machine language by hand, since this was all pretty simple stuff. A big program would be a couple of hundred instructions. At least for *me* that was big. The last step was to calculate the actual memory addresses to be used and lock the program into a particular section of the computer memory. And then, you sat down and put each four hexadecimal digit instruction into the computer memory with the front-panel bit switches, one after another. What a thrill to push the "Run" button and see what happened!

I was interested in running stepping motors and similar things, so I devised a plan to try out some of my ideas. I would build a "handwriting machine," using the computer to store writing motions that could be played back by motors. Two stepping motors and drivers were scrounged, which was easy with nobody guarding the stash of good stuff. I mounted the motors on the back of a panel and built code wheels to fit the shafts. The code wheel was just a hand-sized disk that had a row of holes around the outer edge. A photocell was mounted to look through the holes and generate a series of on and off signals as the disk turned. A simple linkage tied the two motor shafts to a fiber-tipped marker pen. This was all arranged to hold the pen in writing position over a paper tablet.

The plan was to manually push the pen through the motions of writing while the computer stored the movements. Then, with a clean sheet of paper in place and the pen uncapped, the computer would be played back to reproduce the handwriting. It didn't look very pretty, but it worked and had great entertainment value. For example, a person could write his name during the ten seconds that the computer was storing the motions, and then play it back at twice or four times the speed. Boy, could you write fast! The accuracy went down a lot, but it was spectacular to watch if it didn't sling the pen out of the linkage and across the room.

The computer sampling rate that I had put in my program to get reasonable accuracy in duplicating the motion was such that the 4096 word memory was full in about ten seconds. But it proved my point and worked if your name wasn't too long.

Along the same lines, I built a delta modulation system to record speech. Since it ran out of memory even quicker than the writing machine, you spoke into a mike for only a few seconds and it would then play back a barely understandable version of what you'd said. On my first try, I got as far as "What hath Ed..." before it filled up the memory.

But the best was when vacation was over. My old buddy and later boss, Clyde Golightly, came back to work and I couldn't wait. Clyde was way smarter than the average person, and very, very difficult to pull anything on. He could smell a rat a mile off and was rarely, if ever, a victim of the puerile jokes that others, like me, tried to pull on him. But I thought I'd give it another try.

I saw Clyde wandering down the aisle in our open lab area, heading in my direction. He hadn't seen nor heard of my wonderful computerized writing machine, so I thought I should give him a personal demonstration. I quickly set the computer program to the "Record" memory address, then started it. I moved the capped pen and in the air wrote "CLYDE" in large letters. By the time Clyde reached my bench, I had the computer set for the "Playback" mode; the pen uncapped and ready to go.

After a little small talk about vacations I showed Clyde what I'd built. I said, "Clyde, you're gonna really like this thing. It's a computer-controlled voice writer. Watch." I turned to the front of the computer and yelled at it, "Write *Clyde!*" I pushed the "Run" button on the Model Ten and the pen wrote "CLYDE" in an appropriately jerky manner. There was a split second when Clyde was taken in. No more than a hundred milliseconds, but it was enough. Then he realized he'd been had. He walked off, shaking his head, a broken man. Yeah, right.

The next day, Clyde again came walking down the aisle towards my writing contraption. I was ready for him this time. First I apologized to Clyde for being a smart-ass the day before, and told him that I wanted to really show him what I'd done with the Model Ten, and that it was pretty keen. He wanted to see it work.

I coached him through the steps of writing his name, and he dutifully wrote "CLYDE" on the paper. A clean sheet of paper was installed, and I pushed the "Run" button. The pen wrote "CLOD." Got him again. Of course I'd pre-written "Clod" into the computer and then faked his "Clyde" recording. It was a rare event to zing Golightly once, much less twice. And of course, naturally, I began working for him in 1973. Luckily for me he didn't hold a grudge more than a couple of years. Yeah, right.

About this time, off in a corner somewhere, Ed Jackson was thinking. Ed, as you may recall, was on Paul Davis's team that built the first transistor radio in the world back in 1954, and the solar-powered fan-cooled pith helmet. He was one smart cookie.

I had known Ed from the early 1950's and recognized him as one of the more interesting people at TI. One of his fascinating non-TI sides was his love affair with pipe organs. He and Mac McDonald, maybe the second most interesting person of my acquaintance, personally rejuvenated the old Palace Theater pipe organ in downtown Dallas. The silent movie era organ was a Wurlitzer Model "Publix Number One," with twenty ranks of pipes. Ed and Mac both later bought abandoned pipe organs of their own. Mac's is now in the Lakewood Theater in Dallas, for those of you who like to keep up with such things.

But with the unselfish efforts of Ed and Mac, the Palace organ lived again after being silent for many years. One Saturday morning, Mac and I went down to the Palace before the matinee for a grand tour of the refurbished organ. It was an immense apparatus, as large pipe organs are, with bits and pieces tucked here and there all over the theater. And then, and *then*, I got to ride the console out of the basement and up into the theater while playing *Nearer My God To Thee* on the reiterating xylophone. I had achieved a musical peak of some sort, if not a true climax. Mac let me play it for fifteen or twenty minutes before the Saturday matinee began and it was a blast. It must have been like flying a B-29, what with all the controls and switches and levers. And a B-29 couldn't have given me any more a feeling of power than making the Palace reverberate with that mighty Wurlitzer.

While on the subject of pipe organs, Ed Jackson had a really great idea for a practical joke, but never put it into action. His plan was to attend a wedding (preferably of a friend or acquaintance) that was occurring at a pipe-organ-equipped church. Before the ceremony, he would find the pipe organ blower and fasten one end of a long garden hose to the air inlet of it. The other end of the hose would snake out to his car in the parking lot. In the trunk of his car would be a tank of helium.

As the organ broke into the stirring strains of "Here Comes the Bride," Ed would open the valve on the helium tank, flooding the pipe organ blower with the gas. Since the velocity of sound in helium is three times that of air, even a small amount of helium in the air would cause the pipe organ to go horribly out of tune, with the pitch rising as the percentage of helium increased. Ed would then turn off the helium, retrieve his hose, and leave. The pipe organ, if not the bride and groom, would shortly return to normal.

A pipe organ repair person would not believe the story, since there is no way an organ can go out of tune all at once. Unless, of course…

I recall Ed Jackson had designed and built a "Seismic Radio" for GSI in the early '50s that was a sight to behold. It was easily the most complicated radio I, or maybe the world, had ever seen. But the story goes that during the development of it Ed needed to use some higher mathematics that related to calculating electromagnetic propagation. Ed never finished college, probably because he was bored to death, and wasn't familiar with the math.

So Ed looked around and found a graduate class at SMU that was offering this bit of mathematical esoterica and went to sign up. I understand it took some string-pulling by TI to get Ed in, since he didn't have a college degree, much less the required prerequisite math courses.

I was told that after a few classes with the instructor struggling with the material, Ed himself took over the class and taught the remainder of the semester, learning it from the book before each class period. If you don't believe this, you never met Ed.

But Jackson's smarts were not just in book-learning. He was called in as a last resort to see if he could determine why a photolithography process in one of the integrated circuit production lines worked one day and not the next. The process engineers had plotted out a long and complicated series of tests to see if they could dope out the probable cause, but thought maybe Ed could see something that they had missed. It was worth a shot. After Ed had been brought up to speed on the problem, he asked them to hold off for a day or two while he looked at it and they agreed. Any fresh ideas were welcome.

Ed first got the factory specifications on the light-sensitive photoresist that the process was using, and then drove downtown to the Kodak supply house. After studying the spectral sensitivity curves of the resist, he picked out a suitable Kodak Wratten optical filter and headed back to TI. He walked into the production area, held the filter to his eye and scanned the room. He left and brought the process team leader back with him to the area and pointed to a yellow plastic window that was supposed to shield the photoresist from room light. Ed said, "That's not the right material. Some of the blue spectrum of the outside light is getting through and fogging your wafers."

And of course Ed was right. The Kodak filter Ed purchased only passed the light that exposed the photoresist. In the production area, he held it to his eye and looked around until he saw light, and there was the cause of the problem.

As I was saying, about this time Ed Jackson was sitting somewhere, wearing his signature sandals and white socks, and thinking. And what he was thinking about was how to build integrated circuits. The current production methods had growed like Topsy, without any overall authority, each process independent of the other. The silicon wafers were carried by hand from process to process, and dipped, dunked, rinsed, coated, stripped and subjected to various other indignities, until finally, with luck, they contained a few good integrated circuits. Ed thought this really sucked and was such a disorganized approach that the yields, i.e. the percentage of good devices, would never be very high, and thus the cost of the ICs would come down slowly if at all. And Ed was right, through and through.

Ed, by himself, which is how Ed worked best, began to devise a true manufacturing system where the silicon wafers would be passed, one at a time, automatically from process to process without human intervention. And like most pioneers, the processes weren't really ready for such organization. Several of the process steps were done in "batches," not one at a time, and most of the steps required some human skill to perform. But Ed knew what computers could do, even at this embryonic stage of development, and decided that computer control was just what the world of IC manufacturing needed. Had I known what he was working on, I would have agreed wholeheartedly.

Ed's first poke at the hardware was to find a way to move the silicon wafers from one process station to the next neatly and cleanly. He found a way that was patented by another company and investigated it. Dr. Warren Rice, once again, was brought in to take a look at this "air track" concept. It looked like just what the Doctor (Rice) and the non-doctor (Ed) ordered. TI bought a license to use "air track" internally on process equipment. It was a good decision.

Ed Jackson teamed up with Jim Nygaard to present the idea of an automated integrated circuit factory to management, especially Pat Haggerty. Pat, the President of TI and the foresighted executive who literally forced Texas Instruments to get into the new-fangled transistor business in 1952, would be the one to say yes or no on a project of this magnitude. And it would be a biggie. Virtually nothing existed that could be used in it. Everything would have to be designed and built from scratch. And it would take a lot of scratch to build an automated factory from a technical standing start. Jim and Ed's enthusiasm and belief in the idea must have been contagious because Haggerty bought it and funded it.

161

Thus, the Front End Project, more commonly known as the FEP, was born and rumbled into action. It was going to be designed to process the latest and greatest silicon wafers, which were a whopping two inches in diameter. None of the small stuff for this baby. A team was pulled together to design and build a prototype.

My involvement in the FEP was limited to some minor work on the computer control system and in the design of the machine interface boxes that received the computer signals. I recall one specific criterion for our control boxes: the circuitry had to check the format and content of the incoming computer command for validity. What if the computer messed up and sent the wrong signal to our machine? It could be hazardous, what with it controlling chemical valves and such. We really didn't trust the computer and the penalty for a misstep could be great. The FEP control system was going to be equipped with a belt *and* suspenders.

Meanwhile, back in my area, we had a brainchild of our own moving to the front burner. After programming and playing with the Model Ten, two things became apparent: first, computers make swell controllers for equipment, and second, the way the Model Ten had to be programmed wasn't the way to do it.

The Model Ten was a take-off of the IBM 1130, and it was designed to handle numbers. But we weren't interested in numbers and wanted to be able to manipulate ones and zeros, or bits. One "word" in the IBM 1130 or the Model Ten consisted of sixteen bits. To us that was sixteen motors or lights or valves that we could turn on or off. And another sixteen bits could be sixteen different signals from the machine, like the position of a switch, or a float level, or whether a photocell could see a silicon wafer or not. Bits were what we wanted to deal with, not numbers. Production machines were controlled by bits.

So my group decided to build our own process control computer. A computer that would be designed to deal with bits, not words, and designed to interface with the world of motors and switches. We wanted a *bit-pusher* computer, not a number diddler. The first person I collared was Lee Blanton. Every new design needed Lee on it and this one *really* needed Lee.

We spent several weeks developing the format of the computer "word" and the instruction set, with Lee being the leader of the group. In the midst of a heated discussion about the amount of memory that the machine should be able to address directly, I made a statement that has since been repeated back to me more times than I care to mention. I said, "Let's go with twelve-bit addressing. Nobody in their right mind will ever need more than four thousand words of memory." And people say I lack foresight...

I carried the idea to my new boss, Don Walker. Don, who had properly come up through the ranks at TI, grasped the idea immediately, and also grasped the idea that bootlegging it would be the only way to get it done. He quite rightly surmised that a very low profile would be needed. Others in TI were supposed to be designing computers, not us, and jealousy can be an ugly thing. Particularly jealousy across TI Division lines. So Don arranged to snitch money from other projects and bootleg the development of the HAL-9 computer while conveniently looking the other way and whistling nonchalantly.

As you might have guessed, we had some fun picking out the name for it. Since the movie *2001,* just out in 1968, had the world's most powerful computer, the HAL-9000, it seemed like whipping up a tiny version of that would be a good thing.

After the instruction set was nailed down, we turned Mike Bunyard loose on the circuit design. Mike, a West-Texan and a Texas Tech graduate, apparently was also born knowing how to build computers. Mike completed the first-pass logic design in one week flat, using the standard TI fourteen-pin DIP (Dual Inline Package) TTL (Transistor-Transistor Logic) integrated circuit family. In four weeks he had cycled the design around twice more to reduce the package count to a manageable size and commenced the printed circuit board layout. The prototype, complete with a snazzy plug-in front panel, worked almost off Mike's drawing board. It was an awesome piece of engineering skill. And I'm proud that I was able to give Mike the splendid leadership he needed to accomplish it.

With the prototype running nicely, we shopped the HAL-9 around the Mechanization Department that was building production equipment and immediately hit an interested group. They had set up, once again, a project team to attempt to build an automated version of the hand-operated wire bonder. This device was used to connect the tiny computer chip circuit to the wires on the package. It was an incredibly difficult task, and other groups had failed previously. But the idea of each bonding machine having its own computer seemed

163

like a real leg up in controlling the complicated system. In the spring of 1969, we began building HAL-9s in earnest.

It was during this time that I was teaching a short course in programming the HAL to interested engineers, most of who had never been near a computer. It was a short course because it just didn't take that long to learn "how to." A couple of hours of instruction and a person was off and running. That was another advantage of the HAL—people could understand it. This carried over into the HAL-9 assembler that some smart person, I think either Blanton or Rylander, had written on the IBM 1130. Put in HAL-9 assembly language and it would spit out a paper tape with machine code. Saved a lot of time and was more accurate than doing it by hand, no matter how much fun I thought it was. And the assembler had some "Do What I Mean" commands in it. In places where there were no conflicts of interest, the assembler would accept an Oh in place of a Zero, or vice versa, and the same for Ones and Eyes. We trusted that some day the neophyte programmer would learn the difference and that punishment was uncalled for at this time.

I had a class of three or four engineers and was demonstrating the "Add" command on the HAL by writing a tiny machine language program to perform the mathematical function, "2+2." With the group watching the binary "bit" lights on the front of the HAL, I pushed the "Run" button. The binary answer, 0000 0000 0000 0100, which translates to 0004 in hexadecimal and just plain "4" in decimal, showed up the instant I pressed the button. One of the engineers whistled in amazement and said, "Boy! That was fast!" He just didn't have the concept of a 3.5 microsecond add time. It was a new world.

If you wanted to buy a HAL-9 computer, it cost you $5,000 for the Central Processing Unit, which was just the HAL-9. Now if you wanted memory, like you wanted to actually run it, you needed to buy an Ampex RF-4 core memory to go with it. That sucker cost nearly $7,000.

The HAL-9 was a five and a quarter-inch tall standard nineteen-inch wide rack-mount unit, while the Ampex RF-4 required two of the five and a quarter-inch boxes. Put them all in a tabletop equipment rack and King Kong himself couldn't lift it off the table. We used to tell people to "clear core," i.e. put all zeroes in the memory, before moving it so it would weigh less. Some of them did, but they still got hernias.

Near as I can tell, we eventually built about fifty HAL-9s and none were sold outside TI. They were used in a variety of systems, with thirteen of them in the first successful automated wire-bonding machine, the ABACUS. The automated wire bonder project had finally struck gold with the help of the HAL-9 and Mac McDonald, the crazed engineer turned crazed software person. Mac not only learned how to use the HAL-9 efficiently, he taught it to do things it wasn't designed for. Mac was able to get around the shortcoming of the limited instruction set and make the ABACUS sit up and bark. It was a fortuitous joining of forces.

This first ABACUS with the HAL-9 was the beginning of a long line of distinguished ABACUS models that ramped-up TI's integrated circuit production worldwide and dramatically lowered the manufacturing costs. The ABACUS II, actually the fourth design, was awarded The American Society of Mechanical Engineers' "International Historic Mechanical Engineering Landmark" designation in 1992. It used the TI 960A computer, and over a thousand were built. I like to think it was because we got them started in the right direction.

But TI management finally caught on, alerted by our ABACUS success, and we were told to cease and desist building HAL-9s. TI had announced the 960 and 980 computers and the official word was out to use them in all new projects. This was the right corporate decision since the 960 was a splendid machine-control computer and much more powerful than the HAL-9. But we made our point and learned a lot about computers in the process. Don Walker had done the right thing also. As a kind of grand finale, my personal HAL-9, which was given to me (legally by Jim Moreland) on my 20th anniversary with TI, eventually ended up in the Smithsonian.

But since we had the hang of it, we ("we" meaning Mike Bunyard) designed and build two more little dinky computers to control small machines that couldn't afford a real computer. A number of HAL-009s and HAL-004s were built and used in simple production machines. As you can imagine by the name, they were a subset of the HAL-9 and very limited in horsepower. They had enough power in low gear to get over small buildings, but not in a single bound.

To put an end to the HAL line of computers, I conjured up the HAL-001, the world's weeniest computer. It had a one-bit operation code and theoretically could actually perform useful functions. But not very many. It remained a paper exercise.

We had begun to build machines run by computers and I had developed the grand opinion that we were pretty darn smart. The computer world was our oyster and bring on the challenges. We could do just about anything, given a pad of coding sheets. And then Ed Jackson, who heard me expounding in this self-congratulatory mode, asked me a question that I've never gotten over. "Ed," he said, "Would you write me the program for an ant?" That kind of put our work into perspective. We hadn't even begun.

As the cool and invigorating weather of fall arrived, I began getting that restless feeling again. It had been an incredibly rewarding year, but somehow I had the "rut" syndrome rising within me. Adding fuel to the fire was my dear friend Steve Karnavas. Now the boss of Textool Products, Steve was making attractive noises in my direction. Quit Texas Instruments and go with a whole different company? Why, sure!

My accrued vacation carried my official termination date into January of 1970, but the fifth of December was my last day of work at TI. Onward to Textool!

Chapter Eight

Textool Products Incorporated was located in Irving, Texas, which at the time was a reasonable drive from our house in Lake Highlands. It took thirty or forty minutes to loop around LBJ Freeway and across to Highway 183 and cut down to Pioneer Drive. Don't bother trying it today. Today it would take a helicopter to equal my 1970 driving time.

It was a modern brick building with offices in the two-storied front, engineering in the center and the machine shop, plastic molding, and assembly areas in the rear. Parking was alongside the building and about fifty times closer than I was used to at TI.

They still did general machine work for various customers, but it was a kind of leftover business from the earlier days. They also did custom plastic molding, including building the complicated mold tooling. The shop was equipped with a nice collection of plastic molding machines of various sizes, up to a machine that could shoot a plastic part weighing four or five ounces. They specialized in small, precision plastic parts, not the big stuff.

But the main moneymaker was the line of sockets that Textool built for the electronics industry. They made dozens and dozens of types of sockets for transistors and integrated circuits. In addition, if you had something weird you wanted to plug in, Textool would build a custom socket for you.

Off in one corner of the Textool building was the graphite shop. This well-paying niche business served the transistor manufacturers. The graphite shop machined "graphite boats." A graphite boat was a precision-machined block of graphite. Graphite is like writing pencil lead only a lot bigger. The boats were used to hold tiny parts of transistors in position while heating them in furnaces to hundreds of degrees Celsius, like red hot. A graphite boat was basically just a piece of black graphite with a lot of holes drilled in it. It could be easily held in your hand and looked like an easy piece of work. Wrong. The accuracy of the hole sizes and the spacing between the holes was measured in tenths of thousandths of an inch. The customers paid dearly for graphite boats, and Textool earned every penny they were paid. Few companies could make them at all and none better than Textool.

Sam Braun, the German machinist with whom I had the
pleasure of working in the very early transistor days at TI, started
Textool in the late 1950s. Later, Textool and Sam were instrumental
in getting our CAT machines built in a timely fashion.

Sam kept an eye open for business opportunities and began
making a few graphite parts for TI. He recognized a growing
opportunity, especially if he could build them better than anyone
else.

So Sam designed a unique pantograph especially to machine
graphite boats and it was far superior to anything available. Textool
built themselves a few of these peerless tools and then began
cranking out high-quality graphite boats. And to make sure his
quality was better than anyone else's, Sam added an additional
requirement. Other vendors of the boats would measure their
finished products and "scratch out" any holes that weren't within
specifications. The customers would buy these parts at a reduced
price and use only the holes that were correct. Sam had a
revolutionary idea—Textool wouldn't ship graphite parts that had
even one bad hole. Only perfect boats would go out the door. He
reasoned that if his workers knew that they could sell faulty parts,
they wouldn't be as careful when they were making them.

TI was aware of Sam's quality control dictum and bought all
the graphite boats Textool could turn out. But one day, TI needed
more than Textool could produce and came to Sam with a proposal:
"Sell us your reject boats that have defective holes. We don't care
and we'll even pay your full price. We're in a production bind and
need to get more boats in the furnaces." Sam's response was to go to
the cabinet where the reject TI boats were stored, take them out and
saw them in half on the band saw. No, repeat, *NO* defective parts
were going out of Textool. It wasn't a matter of money with Sam, it
was his lifelong pride of workmanship.

A few years before when I was working in Science Services
and building the digital seismograph, I nearly got the shop foreman
fired at Textool. He had called me about an aluminum plate he was
machining for me to say he had misread my sketch and drilled a hole
in the wrong place. He knew I was in a hurry, but he would have a
new part built as quickly as he could. I told him to plug the hole and

that it really didn't make any difference on this particular piece. It was not a problem.

But it was a big problem when Sam spotted the plugged hole when I came out to pick it the plate. I had to do some fast and serious talking with Sam. The foreman had done what I'd asked, but Sam was adamant about poor workmanship. If Sam hadn't known me as well as he did, he would have fired his foreman for plugging the hole.

The graphite shop was a walled-off and separate entity from the regular machine shop and assembly areas, and it was a good thing. Machining graphite fills the air with a black carbon-dust cloud that permeates every crevice. Each of the pantographs used to shape the parts was equipped with an exhaust system to capture the dust, as were all the saws and other power tools. And it almost worked. The area was remarkably clean considering the opportunities for turning black. But the carbon dust would creep into the electric motors and with regularity, a motor would emit a large flash of fire, a small mushroom cloud of smoke and cease operating. Since my job at Textool was not very well defined, I shouldn't have been surprised to find that I, the only electrical engineer, got to change out the motors on the pantographs. It figured, I was told, since the motors and I were both electrical.

Sam Braun was President of Textool and his son. Fred, was Vice President. Fred, whom I had also known at TI, was a gifted machinist and toolmaker. He much preferred turning the cranks on a Bridgeport mill to doing vice-presidential things. I rarely saw Fred upstairs in his office but often in the shop.

By the time I went to work for Textool. Sam was semi-retired and working part time. I didn't see much of him, but when I did he was always gracious and pleasant. He had done a terrific job in building Textool to its present state from a garage machine shop. Sam had plenty of talents away from the milling machine.

The General Manager and mainspring of Textool however, was Steve Karnavas, ex-TI transistor manufacturing manager and a lot of other things,

Steve, without much visible effort and in a laid-back manner. managed Textool neatly and in a thoroughly business-like manner. It was well organized and efficient. It was a nice change from the loosey-goosey style that I knew and loved at TI.

My main task at Textool was to manage the Engineering Department and design sockets. My job came well equipped with people who fortunately knew a lot more about sockets and plastic molding than I did. My main man on the drawing board was Phil Brookshire. Phil had a talented pencil and was especially good at designing the complex molds for making the plastic parts. His job was 50 % science and 50 % art.

I also had a hearing-impaired designer. He also was a crackerjack mold designer and every manager's dream employee. Nothing distracted him while he was working. In fact, when the coffee break buzzer sounded, Phil would reach over and jiggle the clothesline along the wall that held drawings and notes to get his attention. Phil then gave the pseudo-American Sign Language signals of a coffee-drinking motion followed by the action of breaking a stick, just in case he didn't know what time it was.

We cycled RFQs—Requests For Quotations—through Engineering by the bushel. There are a lot of people who wanted free estimates on a job, and a few people that would really like you to build something for them. It's not easy to tell them apart, so you treated them all the same.

Many of the requests were for modifications of an existing socket. Phil or I would sketch out an idea for doing it and then we'd pass the information to Steve Karnavas for final pricing.

Requests for new things were more interesting. Phil or I, or both of us working together, would come up with an idea for making it. This took longer and was more difficult to estimate, but Karnavas was good at figuring out what it would cost to do something, even if we weren't. Wisely, Steve didn't let us mess with the money numbers.

It was a tough year at Textool. It was also a tough year at TI and a lot of other places. I had to lay off several people from the Textool Engineering group, and TI also had to lay off people during 1970. Business had slowed down and sales of sockets and graphite were weak. We scrambled for every opportunity and didn't dismiss any inquiries. The next one might be the big one. We did a lot of estimates for customers and prospective customers and enough of them were successful for Textool to survive.

In the fall of 1970, I stayed late one evening trying to work through some quotes and check out a new socket. I was the only person in the plant that night and it was nice and quiet and the phone didn't ring. I could get a lot done at night.

The new socket was for a 24-pin DIP integrated circuit. The DIP units were hard to push into the sockets even with their usual 14 or 16 pins, but the 24-pin integrated circuit was really tough. Especially when you're trying to plug it in for the first time, before the contacts get "broken in."

I put the socket down on my tabletop and lined up the DIP on top of it. I pushed the DIP down firmly with my thumb but nothing happened, so I pushed harder. Then something happened—the DIP unit flipped over, upside down, and embedded a number of its pins in my thumb. I'm sorry to say that this was a fairly common accident when trying to plug DIP units into sockets, and it really seemed kind of stupid.

I pulled it out of my thumb and bled on my desk while I said a few choice words. Maybe we could generate a new old saying like, "Bodily injury is the Mother of Invention." Without going into any imaginary details about how an invention sprung forth full-blown from my brow, let me just say that, by golly, I did have an idea. It was the old light-bulb thing again. What the world, and especially Textool, needed was a socket that you could drop a DIP integrated circuit into easily and then somehow clamp the contacts afterwards. I set aside the pile of overdue RFQs, wrapped a handkerchief around my thumb, and got out a clean sheet of paper. I had finished eating my spinach and it was time to have some *fun!*

For the next few days I sketched and made trial parts to test things, and pretty well got a final concept out of my brain and onto paper. Except for one lousy detail. The center plate in a three-plate plastic sandwich had to be moved with some force about an eighth of an inch to properly lock the contacts. I had sketches of ideas all over my desk and none of them were any good. My brain had gone dry. Then I had one last idea—I stuck my head out in the hall and called to Darien Fenn, an interesting young man working on the drawing board.

Darien was the son of Syd Fenn, a mechanical engineer friend at TI. But Darien couldn't seem to get through engineering school and had been treading water in his life while working as a draftsman. I liked Darien a lot and it was obvious that he was smarter than old Billy and just hadn't found the right direction.

When he came in I sat him down and explained what I wanted from him. "Darien," I said, "I'm getting ready to leave for lunch and I have a little job for you while I'm gone." I showed him my quandary with the three-plate sandwich and said, "I want you to come up with twelve ideas on how to move and lock that center plate. One dozen. By the time I get back." I left for lunch.

When I got back from lunch there was a sheaf of papers on my desk. I counted the sheets and there were twelve of them, each with the doodlings of a mechanism. And on one of them was the sketch of a piece of cleverly bent wire. That was what I was looking for. I couldn't come up with it myself, but I knew it when I saw it. It was perfect. It cost about a dime and couldn't miss. The kid was a genius.

This exercise turned into the Textool "ZIP-DIP" line of sockets which are still being produced. "ZIP" is the acronym for Zero Insertion Pressure, and by this time you know what DIP stands for. Textool sold a gazillion of them in all sizes.

I still have a copy of my check from Textool as payment for the patent. I split my reward with Darien for his help, so he got fifty cents. He also thought up the name "ZIP-DIP" but didn't get any more money for that. You don't want to overdo that sort of thing.

For some closure on Darien, he ended up as Doctor Fenn, with a doctorate in psychology. He posted a 4.0 GPA as he went through college. He may not have wanted to be an engineer, but he was.

But the ZIP-DIP was my swan song at Textool. I really missed the energy and excitement of the ragged-edge technology that I'd been working with at TI.

By this time I knew the drill. I called Jim Nygaard and asked him if all was forgiven and could he fit me into his group at TI? Yes, he said, he could. I hated to tell Mary Ruth.

Chapter Nine

Jim Nygaard must have called in some markers to get me hired back into TI on the last day of 1970. When I checked into the Personnel Office to be "processed," which sounds like what they do to beef, they couldn't find the forms for hiring people. Lots of forms for laying off, but none for hiring. I was told that I was the first person they'd hired in a long time. But they managed to get some sort of paperwork into the system to accept my employment and my first day of work was to be the 31st of December, a Thursday.

I made a special effort to get back on the TI roster in the year of 1970, the same year I left, because of some of the peculiarities of the Profit Sharing, vacation benefits, and seniority calculations. It seemed like it would be worthwhile to arrive on the last day of 1970 instead of the first day of 1971. One never knows, and I'd heard horror stories of missing a year's profit sharing by being on the wrong side of the year boundary. Besides, I knew I wasn't going to work very hard on the day of New Year's Eve, and my second day would be a holiday. It would be a good way to ease back into working at TI.

Jim Nygaard was on vacation, along with almost everyone else at TI when I arrived for my first day of "work." The great majority of the people with any seniority at all took the week between Christmas and New Years off since it was a nice time to be with the family, and things were pretty quiet around TI anyway. So I found where my desk was, and spent the day wandering around the South Building getting reacquainted with a few people and trying to find where things were.

My new location was in a small area that included Jim Nygaard and his conference room just outside the FEP "Front End Project" manufacturing facility. This is the giant project conjured by Ed Jackson that had been started in 1968 and put into production in 1970 while I was away at Textool. It was a hell of a bunch of machinery.

The interesting part of the facility was a large room with six automated production lines. Each line performed the

173

photolithography for one "process level" on a silicon wafer. This meant that the wafer was coated with a light-sensitive "photoresist" material, baked just right and then passed into an "Align and Expose" machine. This machine, with the guidance of an operator, aligned patterns on the surface of the wafer and exposed it. Following was a process station that did the "develop" step, and then passed the wafer on for more baking. It was basically a photographic process and a lot like what happens when you snap a picture and take your film to the drugstore. The wafers coming off the line were then sent generally to another room for additional abuse, like diffusion or etching, although a couple of the automated lines also did the etching. The wafers were then stripped of the used photoresist and started in another FEP line for the next process level. The wafer needed five or six levels of processing to make an integrated circuit, and so had to pass through all the FEP lines sequentially.

Computers controlled the whole room of equipment. It was way ahead of its time and by virtue of the talented groups who built it under Jim Nygaard's leadership, Herby Locke's inspired running of it, and Don Benefiel and Bob Falt's process engineering magic, it was a winner. A year or two later it set a record that has never been equaled in a "Front End" at TI—it processed over 100,000 wafers in one month. Granted, the wafers were two-inch wafers and the circuits were much simpler and required fewer levels of processing, but it was still one hell of a production rate. The FEP production personnel were probably inspired by the ubiquitous signs Herby Locke put up in the FEP area, "Up Your Yield!" The proper response was, "Well, up yours, too, Herby!"

My desk, along with two others, was just outside the door of Jim Nygaard's office. Behind me was Ed Jackson, all-around scientist, and ahead of me was Ed Ducayet, bean counter, a.k.a. accountant. The state of the three desks looked like a linear progression into a disaster. Ed Ducayet's desk was absolutely spotless. Nothing would dare remain on the top overnight. Mine, in the middle, was a reasonable clutter with parts and piles of papers, and well within what I thought was acceptable housekeeping for an engineer. But behind me, Ed Jackson's desk look like a hazardous waste dump and lacked only a flag with skull and crossbones on it.

Ed Jackson was a lot more interested in things he was doing than in the state of his desk.

So guess whose desk the night cleaning crew ate their lunches on. Of course it was the clean one. Ed Ducayet would become absolutely livid in the mornings when he would find crumbs, mayonnaise stains and drink rings on his formerly pristine desktop. I never had the problem, and the crew couldn't find a level spot on Ed Jackson's desk. Ducayet finally got Nygaard to spring for a spray can of powerful cleaner and disinfectant out of petty cash, but he was never able to stop the third-shift picnicking.

One of the first things I got to do after I hired back in was to travel with Nygaard to Nice, France and Bedford, England. Both places had TI semiconductor manufacturing facilities and things of great interest for us, but for the life of me I can't remember what they were. I had never been overseas before and thoroughly enjoyed seeing Nice from the balcony of my room in the Hotel Negresco, and the sand and surf of the Riviera of France. TI sure knew where to build transistors, all right.

The two stories I recall about the first year or two of the Nice plant may be apocryphal but I like them. The first story was the amazement of the personnel department at how easy it was to hire all the workers they wanted at reasonable wages. One of the concerns had been that getting people to work on the Côte d'Azur would be difficult if not impossible, and that the wages would be necessarily high. And just the opposite was true. At least until the first warm and sunny day in the spring, when only about a third of the employees showed up for work.

The second story relates to the difficulties in procuring the chemicals that were needed to manufacture transistors. The French government was adamant about keeping suppliers in France, so it had been a scramble to find suitable producers of the high-purity reagents and solvents. The story goes that one supplier was grossly overdue in shipping needed supplies, and a buyer from the Nice plant went personally to the small chemical company to see what the problem was. The problem, he was told with a Gallic shrug, was that they were not able to procure enough empty wine bottles to ship the product in.

The highlight of my visit to Nice was an evening spent in a tiny walled medieval village a short distance from town and dining there in an old and elegant restaurant. It was memorable because during the meal Jim Nygaard excused himself from the table and asked the TI Nice accounting department manager, an Englishman, where the men's room was. He was told without hesitation, "Out the front, up the stairs to your left, and it's the door at the top."

I commented to the Englishman, "You must have eaten here before."

He replied, "No, and I have no idea where the loo is."

A few minutes later Nygaard reappeared, looking slightly wide-eyed. "I followed your directions and when I opened the door, there was a French couple sitting on a couch watching television!"

Jim and I hopped a British Airways *Trident* and flew to London where an engineer from the TI Bedford plant met us. The drive to Bedford was uneventful, although driving 90 miles per hour on the wrong side of a freeway took some readjustment of the safety circuits in my brain.

We stayed at the Red Lion hotel in beautiful downtown Bedford. The notable feature of our accommodations was that we had one of only two rooms with a private bath. Some days you just luck out.

But the trip to England and Bedford was entertaining and enlightening and thoroughly enjoyable. I just have no recollection of what we did at the plant, although I'm sure it was very helpful for TI.

And so, refreshed by foreign travel, it was back to the FEP and reality. My job was to design upgrades for the equipment and to fix any niggling machinery problems. I began getting into the swing of things by regularly attending the "90% Yield Team" meeting twice a week at 11:00 am in Nygaard's conference room.

The 90% Yield Team was a group of scientists pulled together from the TI Central Research Lab and the Semiconductor Research and Development Lab to analyze the product failures coming out of the FEP and to devise means of eliminating them. I don't think anyone ever thought we'd reach a product yield of ninety percent, but it was an admirable goal. It was a talented bunch from a variety of disciplines, and included, more or less, Tom Hartman, Dr.

Vernon Porter, Dr. Ron Cox, Clif Penn, Dr. Al Tasch, Dr. Dennis Buss, Dr. Wally Schroen, Dr. Dave Cole, Sam Shortes and occasionally Ed Jackson. I was definitely low man on the technical totem pole.

I still remember smarting from a comment one of them made after I had gotten Nygaard to spring for a new copy of the "Handbook of Chemistry and Physics" for me. "What do you want that for? It doesn't have any pictures." Now that's rude. When I recently asked Clif Penn if he was the one who told me that, he replied, "No, but I wish I had." It just goes on and on.

My attendance in the twice-weekly Yield Team meetings was required but limited to my discussion of what I was working on. It was important, at least to me, but only amounted to about 2 percent of the agenda. But I had to attend 100 percent of the meeting.

The year before, when I was at Textool, Mary Ruth and I had purchased our plot of pines in East Texas and I was now contemplating building a family cabin on it. And as an engineer I was constitutionally unable to use someone else's design. The 90% Yield Team meeting was a godsend. I could handle my 2 percent of the technical agenda with no problem and use the other 98 percent of the time to design a cabin for East Texas. And the good news was that my regular seat was next to Tom Hartman, who was smarter than the average guy by a lot. He was to become my architectural critic.

We worked quietly without attracting Nygaard's attention, and after I had been through some interesting cabin design iterations, such as the Sewer-Pipe House, the Ever-Upwardly Spiraling Abode, and the Chiefly Clerestory Cabin, the Millis Exoskeletal Structure was chosen. It was swell, although I had been leaning towards the Sewer Pipe House. I thought I could save some money buying used sewer pipe, but it was not to be.

And, over the next few years, the Millis Exoskeletal Structure was indeed built in East Texas by my hands and those of my friends, and enjoyed by all. It pays to have a scientist looking over your shoulder on important decisions. Tom never did like the sewer pipe design. Too conventional, he said.

About the time Jim Nygaard was meeting incredible technical challenges, he was having a non-technical problem that he was unable to solve. He had fruit flies in his office. Not just one or two, but swarms of fruit flies. Since his office was directly below the South Building cafeteria he could make a good guess where they came from, but Jim had called the Facilities people several times to no avail. Whatever they treated his office with was useless. And then an FEP technician solved the problem.

After hearing Jim complain multiple times about his bug problem, the technician presented Jim with a "Fruit-Fly Eliminator." It was a simple device, as most good inventions are.

A spray can of clear Krylon lacquer was attached to one end of a stick about a foot long and pointed towards a Zippo lighter fastened to the other end. To operate, the Zippo was opened and flicked afire, and then, carefully tracking the flight of the offending fruit fly, the button on the Krylon can was pressed. The resulting watermelon-sized ball of flame vaporized the fruit fly and everything else in the lethal radius. Jim loved to demonstrate it, and sure enough, the fruit fly problem was solved. Please don't build one of these.

Working on the FEP was a blast. Or at least it could be if you weren't careful. But the FEP was well designed and ran safely and only had one small "blast" to my knowledge. A hose carrying alcohol failed in one of the FEP modules and sprayed the flammable liquid over the top of the unit. A line operator spotted the trouble immediately and hit "Big Red," the emergency stop button that each module had in a prominent location. The spark from the electrical switch in Big Red set the alcohol on fire, which pretty well fried the FEP module but it didn't spread and did no damage to the room. Yes, we fixed the problem with sparks in Big Red.

Working on the FEP was a blast in the other sense also. It was a marvelous and complicated collection of equipment and the most interesting part consisted of the six wafer processing lines. A line held a collection twenty or thirty process modules, each of which performed its particular magic on the wafer as it passed through. The system was totally flexible and any process module could be moved to any location in the serial-flow line.

The lines were built on a "spine" concept. A waist-high spine ran the length of each process line and the top was fitted with an array of pins and holes that allowed any of the process modules, regardless of size, to be literally hung from either side. The module could be hooked up to chemicals, gases and power and be ready to run the process within a matter of minutes. This effectively stopped the critics' complaint that a failure in the middle of a serial-flow process machine would stop the entire line until it was repaired. The failure of a module in the FEP could be overcome by changing it out for a spare in ten minutes, tops.

I found out years later that the TI architect, Richard Colley, had seen the FEP line and was mightily impressed with its basic philosophy. He was a friend of Jim Nygaard's and a fellow Aggie, and they spent a lot of time talking about the concept and the development of the FEP system. And then Colley designed the Lubbock semiconductor plant on the same principle—a long "spine" or wide hallway with facilities running the length, and "machines" or buildings in the case of the Lubbock plant, attached at intervals along both sides, tapping into the wiring, piping and the central hall. Pretty neat.

One of the strangely critical processes performed on the wafer would seem to be the simplest—baking it to dry the light-sensitive photoresist. It had to be dried just right—not too fast and not too slow. Half-fast was just right. But it was hard to dry it just right because the FEP wanted to run faster than half-fast. So I was engaged to check out a new rumor we'd heard that microwave ovens were being used at another company to bake the photoresist and it was a lot faster and better. So, working with the Central Research Lab guys, I built a microwave wafer bake station for the FEP.

It soon became obvious that microwave baking sucked big-time. The photoresist itself did not heat up from microwaves, being of great and high resistance, but the wafer did, in spades. However, the wafer heating depended largely on the resistivity of the base silicon. Low resistivity wafers would heat up and fry the resist while high resistivity ones would scarcely get warm under the same conditions. Control of the baking temperature, at least with the technology available at the time, was impossible. It was a failure but I mention it now because microwave baking, like the proverbial cat, had at least nine lives. Every four or five years, someone would reinvent microwave baking.

The story was always the same—rumors would allege that some semiconductor company was using microwave baking and it was terrific. The result of the rumor was always the same, too. TI would try it again and it would fail again. Maybe the other companies were starting rumors to make us use up our development money.

Just for calibration, in this year of 2000, the specification for baking a wafer "properly" requires that the uniformity of the temperature across the diameter of a 300-mm (12-inch) wafer must be within 0.25 degrees Celsius. Children, don't bother trying this in your microwave at home. It won't work.

Working with the scientists from the Central Research Lab was a lot of fun. I had a chance to work with Sam Shortes during this time. He was a photolithography and everything else expert as well as being a top-notch artist. I worked with him on a project to develop a method of stripping the used photoresist from the surface of the wafer. Typically, nasty chemicals were used to eat the photoresist off, which worked well, but it was messy and dangerous. I wanted to try a direct approach and strip it off with a high-pressure blast of water.

I built a test rig with an air-powered hydraulic pump that would put out water at 22,000 pounds per square inch. I figured that would be enough to start. Using the Elox machine in the shop, which uses electricity to eat holes in things, I made a nozzle with a five thousandths of an inch diameter hole in it. It put out a very fine jet of water that you didn't want to put your finger in front of. That much I knew.

I began at a much lower pressure, like several thousand psi, and tried stripping the paint-like film from a wafer. No effect. Photoresist, when well baked, did what it was supposed to do—it stuck like crazy to the wafer surface. More pressure was needed, so I cranked it up to about ten thou. The result was ragged removal of some resist and not other. Okay, you asked for it.

With the pump turned up all the way, the jet of water easily stripped the photoresist cleanly from the wafer surface. It also stripped the oxide layer and part of the epitaxial material. When the jet was strong enough to strip the film, it was strong enough to destroy the surface of the wafer. But all was not lost. Sam and I observed that this would be a great way to clean a bare wafer surface. We then built a wafer-cleaning machine that used high-pressure super-clean deionized water. We even got a patent on it titled, "Method and Apparatus for Cleaning the Surface of a Semiconductor Slice with a Liquid Spray of De-ionized Water." The title pretty well describes the invention.

We also learned something from it accidentally. Sam noticed during testing that where the fan spray of high-pressure water hit the wafer surface, there was a bluish glow, like in Frankenstein movies. The spray of deionized water striking the wafer surface at 6,000 pounds per square inch caused an electrical corona, like a high-voltage discharge. Not only was it totally unexpected, we even admitted in the patent that its cause was unknown. But we surmised that it really helped a lot in discharging the static potential that held dirt particles to the surface. Huh? Remember, when in doubt, turn the surprising unknown into an advantage.

Nygaard stayed busy dealing with upper management and attending meetings. I was chatting with him as he was getting ready to leave for an important meeting chaired by Pat Haggerty himself, President of TI, and Jim looked worried. He said that in the meeting the week before, Dr. Willis Adcock of the Central Research Lab had been told by Haggerty in no uncertain terms to do a particular task by the next meeting. And when Haggerty said do something, you did it. The "or else" was understood.

Willis was a good friend and Jim was upset because he knew that Willis hadn't done the task. In fact, he hadn't even started on it. The thought of Willis being drawn, quartered, crucified and then chewed out by Haggerty was upsetting. Jim left for the meeting, plainly worried about Willis.

When Jim returned from the meeting he looked me up. He had a big smile on his face. I asked him what happened to Willis and Jim replied, "When Haggerty asked Willis about the status of his 'must-do' project, I cringed. Then Willis answered, 'Pat, the velocity is still zero but the acceleration is tremendous!' and Haggerty laughed so hard Willis got away with it."

Another favorite scientist of mine was Dr. John Fish, also in CRL. He was a chemist and the resident materials expert and I dropped by to talk with him regularly. Choosing materials to use in building equipment for semiconductor processing was a two-way street. Sometimes the process would eat up the material and other times the material would screw up the process. And, if you were really unlucky, the material you chose would screw up the process *while* it was being eaten up. But John helped me through the maze of chemical mysteries to make my machines and the processes both have at least a fighting chance.

On my first visit to John's office in the CRL building, I noticed a large binder on his bookshelf, "THE CHEMISTRY OF SIN." That looked interesting. I didn't know that sin *had* chemistry but maybe John had a hobby. I finally got up the nerve to ask John about it and was sorely disappointed to find out the it wasn't "SIN," but "SiN," or silicon nitride.

One of the continuing problems of the FEP was the photoresist. It had to be coated just exactly right on the surface of the wafer and tiniest speck of dirt would cause a defect. It was expensive. incredibly sticky and gooey, and cleaning the machines that used it, the "spinners," was a monstrous task. Do you remember this limerick?

> *There was a young man from Racine*
> *Who invented a sex machine.*
> *'Twas concave and convex*
> *And would fit either sex,*
> *But oh, what a mess to clean!*

If you would change *sex machine* to *photoresist spinner* and *concave and convex* to *positive and negative* photoresists, the last line would remain exactly the same. Millions of BTUs of elbow grease and thousands of gallons of xylene have been expended cleaning photoresist dispensers. So I thought I ought to try some new ideas.

I designed a system that would use "throw away" stuff and do away with the cleaning entirely. In my proposed system, the photoresist would come in a plastic bag that looked a lot like a plasma bag, which, in fact, it was in my experimental version.

Molded to the bag was a soft plastic tube a couple of feet long, sealed at the end. The resist would be sealed in this hermetic system, safe from contamination by outside influences.

The operator would hang the full baggie on a hook above the coating station and thread the tubing through a peristaltic pump. A peristaltic pump "kneads" the tubing and makes the fluid flow through it without actually touching the liquid—kind of like milking a cow I would imagine. The end of the tubing would be fed to a position above the wafer and the sealed end snipped off. When the machine was told to dispense resist on the wafer, the peristaltic pump would turn a certain number of revolutions, which would pump out a certain volume of resist. To keep it from dripping afterwards, the pump would be turned backwards a few degrees to "sniff" the resist back up in the tubing. When the resist was used up, the empty bag and tubing would be disconnected and thrown in the trash and a new one installed.

I was sitting at my desk talking with Gordon Pollock, another scientist from the Central Research Labs, and he was idly playing with my prototype photoresist-filled plasma bag. It was kind of fun to squeeze and mess with, but Gordon overdid it and split a seam on the bag. The sticky photoresist shot out all over his trousers. As I looked around for a rag, Gordon, who stammered slightly, calmly got up and said, "You'll h-have to excuse me. I n-need to go back into the y-y-yellow room b-before I expose myself." This was a thought worthy of a scientist. If he could get into the yellow-lit "darkroom" before the resist dried and was "exposed" by the room light, it would be easier to clean. That may not be funny to most people but I liked it a lot.

Along with the prize-winning days in the FEP factory were the occasional "yield crap-outs." A yield crap-out was when the percentage of good devices took a sudden and unexpected plunge to an unsatisfactory level. This instantly led to the formation of a Yield Team to spend their time fixing it, instead of going home and sleeping or seeing the family for a week or two.

And just to be fair with the perversity of inanimate objects, there have been, albeit rarely, "crap-ins" where the yields suddenly rose for no apparent reason. Then a team was quickly formed to find out why the yield went up, so they could keep it there. Either way, crap-out or crap-in, you lost sleep.

But at one particular time the FEP had a **CRAP-OUT**, not just a crap-out. The yield of good devices on many wafers dropped to zero and yet on other wafers was still satisfactory. This was an unexpected and highly unlikely scenario since all the wafers supposedly went through exactly the same processing steps and used the same chemicals and gases. But a lot of them were now failing the probe tests. "Autopsies" on the failed parts showed nothing definitive and there was much head scratching followed by rising panic.

And then someone had a great idea. They got the FEP computer experts together on a crash basis and had them take the data of all the processing transactions that were recorded by the computers, and do a regression analysis. This meant, was there anything, *anything*, that occurred in the days of processing that varied at the same time the yield had changed? The yield of each lot of wafers was examined, along with its voluminous file of processing data by a hastily written computer program. In one day the answer literally popped out at them—when the length of time between the end of a particular cleaning step and the beginning of the metal etch process step increased beyond a certain point, the yield went down. If the time was really long, the yield went down a lot. If the time was short, the yield didn't go down. They found the crap-out by mathematics. It was an elegant example of the hidden power of having computers running your production line.

The problem in production had been caused by a minor change in the method of queuing up the lots to be processed at the etch machine. If the clean wafers sat in the room air beyond a certain time a thin oxide would form on the surface. Little details make high yields.

And then I got a fun project. A new model Align and Expose machine had been put on the market by a west coast company, Cobilt, and the FEP folks wanted some on their automated lines. They were better than the older K&E machines that were now being used. Technology was, as usual, galloping along, so it was time to upgrade the FEP.

We needed a good team, so the ever-versatile Harry Waugh was assigned to be the leader and chief brain man. We quickly added Darien Fenn, who by this time had cycled back to TI and was a known contributor of neat ideas. So with these two and a few others the design of the Cobilt A&E interface began.

The interface was needed to make the Cobilt A&E automatically receive wafers from the FEP air tracks, process them, and send them back to the FEP. Although not designed to be controlled externally, the Cobilt machine would have to be taught to jump when the FEP computer said "Frog."

The FEP-adapted Cobilt was pretty neat if I do say so myself. Harry was a good engineer and manager and Darien was a good designer, as I already knew. The unit not only functioned well but looked good. It even had a particle-free polished walnut armrest for the operator.

One of the key features of the new Align and Expose unit was that it could be changed out rapidly if it broke down. As I've said, on a serial-flow line, if one machine quits, they all quit, so it was imperative that a broken machine could be traded out for a working one quickly. No one ever tried to fix anything on the FEP line unless it was an obvious and simple problem—you got the spare machine and swapped out the bad one. Then, at your leisure, you repaired the defective one back in the Repair and Maintenance shop.

The Cobilt A&E had a lot of connections to the FEP, such as electricity, computer connections, compressed air, nitrogen, vacuum and exhaust. If these were the usual connectors and hoses, it would take forever and a screwdriver to change the unit, so we took a different approach. We used "quick-connects" on everything and mounted them all on a plate.

To attach a Cobilt A&E to the FEP spine, the technician would roll it up and engage two large bullet pins on the A&E into two tapered holes in the FEP module. He would then reach down and turn a large crank which would literally screw the two machines firmly together, automatically making all the connections. It took maybe thirty seconds. Darien Fenn promptly dubbed our new machine, "The Crank-In-Spine Monster." I told you he was a genius.

I was asked one afternoon by the integrated circuit Final Test foreman if I had time to explain the virtues of a Millis Handler to an English engineer from the TI Bedford plant. I was glad to oblige, and met the engineer and began chatting with him about their plans. He had the only Cockney accent I'd ever heard in person, and it was a strong one. I could barely understand him. What do they say about

England and America? Two countries separated by a common language?

We stood watching the handler running in production and he asked me a question about the practicalities of operation that I couldn't answer. I asked the older black lady who was tending the machine if she could answer his question. She replied in a strong East Texas accent that she had no idea what he just said. I looked at the Englishman and he looked puzzled and asked me what she'd said. They were unable to communicate. I, on the other hand, could just barely understand them both and was able to translate back and forth until the engineer got his questions answered.

In addition to my small efforts of FEP machine building, another team in the TI Mechanization Group was also working on future improvements for the FEP. One that was dear to Jim Nygaard's heart was the idea of marking each wafer with an individual serial number that could be identified automatically on the FEP line. The Mechanization Group had been working on such a device and had a test model running in the space frame of the South Building. The FEP equipment was seriously Company Trade Secret and any development on it had to be out of the common view, so the wafer marker prototype was running in a small closed room. The marker cut a series of shallow notches in the edge of the wafer that could be sensed by a photocell. Jim was anxious to see it run.

At the appointed time, Jim was ushered into the demonstration room. It was crowded with the proud engineers and technicians that had built the marking machine. After the project engineer described the operation, the machine was turned on.

It neatly took a wafer from the carrier boat, wafted it down the air track and onto a spindle. The wafer was then rotated on the spindle as its edges were notched. The spindle, with the completed wafer, moved slowly to the next transfer position for a handoff, but stopped abruptly in mid-flight. Unfortunately, the driving mechanism beneath the tabletop kept going. And as it kept going, it cocked a powerful spring that should have been moving the spindle. And of course, the spindle freed itself and slammed to its final position, firing the wafer across the room at near sonic velocity. It just missed Nygaard's elbow and shattered into a thousand silicon fragments on the wall behind. End of demonstration.

Nygaard was invited back for a successful demonstration a few days later and dropped by to tell me about it. He said it not only ran just fine, but that they had painted a bull's eye on the wall where the first wafer had hit, just in case. Jim liked that a lot.

During the year I was called upon by Jim Nygaard to switch to my alter ego as a bass fiddle player for an important celebration. We were to play for Mark Shepherd's badge party. The musical group would be Jim on the banjo, Chanslor on the fiddle, Don Walker singing, and me on the bass. We worked up several fairly rude songs with familiar tunes but rejuvenated lyrics, and were to meet in the North Building at the appointed hour to help Mark celebrate. I got my big stand-up bass out of the car and as far as the guard in the lobby.

"You can't bring that thing into the North Building." said the guard, rising from his chair as if to bar the door.

He hadn't reached for his gun, which I thought was a point in my favor, so I protested, "Why not?"

"I'm only supposed to let in visitors and employees. And that thing is neither one," he replied.

"What if I got it a visitor's badge? Couldn't I take it in then? I can sign in for it." I argued.

The guard looked puzzled but realized in some way that I had him there. He was supposed to let visitors in if they had a badge and were accompanied by an employee.

"Well, okay I guess. Sign in for it and be sure and put your employee number and phone by it in case anything happens."

"Well, sure." I said. "That's fair enough." and so I signed in my bass fiddle, "Fiddle, B." as a visitor and clipped the badge on the E string.

We played so well Mark sent us all thank-you messages. I showed mine to my bass.

Meanwhile, back in the ABACUS automatic wire bonder development group, Mac McDonald had been programming up a storm. The later model ABACUS II used the TI 960A computer instead of the barbaric HAL-9, and he had learned how to do absolutely anything and everything with it. But Mac came to me with a serious personal problem. When his wife had a predicament around the house, like the day before when she'd locked her keys in the car, she couldn't reach Mac at TI. This seemed strange since I knew Mac had a desk and a phone and everything, just like a real employee. But Mac had a secret problem—he didn't work at his desk. Instead, he worked in his hide-away in the basement of the South Building.

After swearing a blood oath that I wouldn't squeal on him, I followed Mac down into the basement, taking a circuitous route and checking periodically for people who might be following. Under some pipes and over others, we arrived at a dusty door in a scarcely-lit area that appeared to be on the basement perimeter wall. Mac opened the door with his key and stood aside for me to enter.

I was dazzled. The inside of the room, probably twenty feet square, was brightly lit with a dozen fluorescent fixtures, and sported several office tables and an executive chair of the latest design. All four walls were solid chalkboards except for the door. The boards were filled with computer flow diagrams and strings of code. Mac had built his own damn computer-programming factory. It had everything but a telephone.

This set-up raised a lot of questions for me, such as how the hell did Mac find this room in the first place and how come it's not on anybody's space list? The lights? The furniture? The chalk boards? And after all that, how come he couldn't scrounge a lousy phone of some sort? There were too many questions so I didn't ask any. I told Mac to have Donna call me anytime she needed to reach him and I'd sneak down and pound on the door.

Mac lived there secretly for years, literally, until his group and mine moved to the Floyd Road Building. Twice in the intervening years Donna called me and needed to contact Mac and I slipped down and notified him. Later, he managed a phone line but I didn't ask him how. I rarely understood how Mac did things.

But I must mention, in this illegal and secret hiding place, Earl "Mac" McDonald did the computer coding that made the ABACUS II bonders run, and run to the tune of one of the most successful machines ever designed and built at TI for semiconductor production. I already mentioned that the ABACUS II won an international award for its excellence in design. Well, you can thank Mac for a substantial part of that success. Did I mention that Mac did his best work alone? *Really* alone?

The FEP software group had also been at work adding a neat feature to the FEP production system. The mystique of cost accounting had always bugged Jim Nygaard because it didn't seem to be based on common sense. Take, for example, the problem of determining the value of the inventory in a manufacturing facility such as FEP. This bit of information was needed monthly to close the books. So, once a month, everybody would stop what they were doing and figure out the value of all the wafers that were in the middle of the FEP processing factory. Jim thought it was totally useless time spent and with the subsequent loss of production an utterly ridiculous thing to do. Why should accounting have to close their books every month? Or every year? Or ever, for that matter? Accounting ought to flow along forever, like a river, with trees on either hand, and not be chopped up to suit some stupid custom that was probably a hangover from feudal England.

So Jim inspired his highly competent software team, led by Ken Wickham and populated by Claude Head and others, to do this problem of inventory pricing automatically on the FEP computers. And they did. The only additional data they had to load in the computer was the dollar value of a wafer at every step in the process, and those numbers were available. The computer already knew how many wafers were at each step at any given time. Soon, a mere tap of a computer terminal key would, in a matter of a few seconds, bring up the value of the entire inventory of products in FEP at that instant. It was pretty damn wonderful and the likes of which had never been seen.

But even though Jim's bright idea had saved the shutdown time of the line to price it out, it didn't help the accounting group. Now they couldn't decide when the proper time would be to push the magic button. And sure as hell, nobody wanted to be there at midnight to do it. I honestly don't know how they finally resolved this serious problem. Jim gave accounting a terrific new tool and they tripped over it, at least for a moment.

The FEP manufacturing facility continued expanding and soon our office area was taken over by machinery. I had been shuffled around a couple of times, and I was now the only one left in our original area. I ended up with my desk by the window, looking into the main upstairs aisle, and sharing my space with the blunt end of a new ion implanter.

Ion implanters were the latest thing in 1973. This was one of the first in TI, and half of it was in my so-called office space. It was not a machine to be trifled with. It was the size of an RV and had a couple of hundred thousand volts inside that innocent cover, along with some poisonous chemicals that were used as targets. And I didn't like the idea that the technicians were required to wear safety gear when they came into my office.

By the time they had put up the protective walls around the machine, my entry and egress was through a door that didn't go clear to the floor in the side of a wall. It was like a door on a submarine. I thought it was time to move.

I began my campaign by taking a nice fat Marks-A-Lot and drawing a big square with a dotted line on the floor by my desk and printing inside it, "IN CASE OF EMERGENCY JACKHAMMER THIS OUT AND EXIT RAPIDLY," making sure it was visible from the hall.

Nygaard didn't think it was a damn bit funny and made me clean it off. But I got to move out the next week.

My move was not just to get away from the ion implanter, which would have been a good and sufficient reason, but to start a new project. The two-inch diameter silicon wafer was being superseded by the three-inch wafer and the FEP was on the road to obsolescence. This was to be the story of semiconductor processing equipment until the end of time—the wafers would keep getting bigger and wouldn't work on last year's machinery. This jump from two inch to three inch was not the first. At the time of this writing, the 300mm (approximately twelve-inch) wafer is the hot stuff. They look like dinner plates.

So, on October 18 of 1973, I began work as the Project Engineer on the "Three-inch FEP" development under Clyde Golightly. And once again I bid Jim Nygaard adieu. And once again, it had been a blast.

Chapter Ten

I had rubbed elbows with Clyde Golightly for years and looked forward to working for him. Besides being a good manager, he was a good engineer, and he had a rebellious streak a mile wide. His disregard for TI "authority" was, as Jim Nygaard's, poorly concealed and it made for a freewheeling and interesting operation.

Clyde pulled together a good, big team to do the new three-inch FEP project. I recall a group of about thirty engineers, designers and technicians crammed into our long narrow room. It was perfect—everyone was within an arm's length of everyone else and the interaction between engineers and designers worked well. It was an efficient design effort.

Tony Adams, later to become a Vice President in TI, was the chief mechanical engineer, with MEs and mechanical designers under his wing. Good old Mike Bunyard, the HAL-9 computer whiz, handled the electrical stuff with help by Wayne Sullivan and others. Wayne was a magic electrical technician who could, for some reason, design anything even if he didn't know anything about it. I never did understand where Wayne learned how to do all the stuff he did. He had just returned from working in the TI plant in Munich, Germany, installing and making a transistor manufacturing system run. He and Bunyard made a great team.

Another key player in Tony's mechanical design team was Cecil Davis, brother of Billy Davis that I'd worked with in Science Services. Cecil was hired in January of 1973, and was an outstanding mechanical engineer and later manager. Before he left TI he was elected a Texas Instruments Fellow, an exalted and worthy position indeed.

And a major part of the design of the three-inch FEP, which by now had become the FEP II, was the software. Not only did each separate machine, of which there were about a dozen, require a unique software package, there had to be an overall "control system" for the computer. So for this very large task, I got a large software group—Claude Head. It was a modern-day reenactment of the "one riot, one ranger" scenario. One huge software job? One Claude Head oughta do it.

The team also had Bernie Robertson, experienced mechanical engineer, plus a staff of designers and technicians, such as Lowell

Simmons and Danny Tucker. And of course, to keep me out of trouble, I had Troy Moore working as a Feasibility Shop machinist and technician. We had first worked together on the CAT machines and he was a top-notch hand at designing and building anything.

If I was going to work for Clyde, I first needed a little breaking-in and training. Now that I'd hit the big-time I had to do the department monthly financial forecasts. For my training in this crucial endeavor, I attended a personal session by Clyde himself of what was known locally as "The Clyde Golightly School of Creative Accounting." Every month I would get the actual expenditures from the previous month and then would be required to forecast the expenditures of my group for the remainder of the year. Clyde taught me in a succinct manner which numbers were important and which were not. Some numbers required serious consideration and careful calculation and some didn't. I learned at the foot of the master.

One lasting memory is the computer printout I received every month, starting with the FEP II project in October of 1973 until I retired in the spring of 1989. The column for November was labeled "Bov," like September, October, Bovember... It was a sure-fire way to tell that the printout was authentic.

We had decided to use the successful FEP spine concept for the FEP II, so we divided up the tasks of designing the eleven individual machines that hung from the spine and set to work. We began the designs in the fall of 1973 and installed the first FEP II in the Dallas Semiconductor Building in the summer of 1974.

One little sideline task that occupied a lot of talented time was the computer system to drive the new FEP II. The TI 2540 computer with core memory that each FEP spine used was out of date by this time, but the good news was that the TI 960A with solid-state memory was now available. More bad news was that the IBM 1825 that was used for the overall control of the FEP systems was unavailable, but again the good news that Digital Scientific built an emulator system, the "Meta-4." Get it? "Meta-four, meta-phor?"

And the final part of the computer puzzle was in our hands from the beginning—the RCCA. This stood for "Remote Computer Communication Adapter" and the one and only existing unit had been designed and built by Mike Bunyard, boy computer genius.

This important digital box would allow the Meta-4 and the 960s to talk back and forth on the FEP II, just as it had allowed the Mother IBM 1825 to talk with the TI 2540s on the FEP. Building an RCCA was simple. Mike Bunyard told Wayne Sullivan to build one, and he did. The fact that he didn't have any official drawings, just some of Mike's sketches, and that the motherboard was hand wire-wrapped didn't bother Wayne. He just built it.

To depart from our path for a moment, let me say that under my splendid leadership, we never did get drawings or a printed-circuit motherboard designed for the RCCA. Every RCCA Wayne built was going to be the last one ever built, but it never quite was. Wayne built six more, near as we can count, that were used in FEP II installations in Dallas, Houston, and Lubbock, Texas, and Freising, Germany. They were also installed in the Linear Department and in the Photomask Departments in Dallas. Why spend money on drawings when you're not ever going to build any more? Or, a better question, why spend money when Wayne Sullivan could build them just about as well without any drawings? I would ask Wayne to build "just one last RCCA" and it would magically and effortlessly appear a month or so later. At least it was effortless on my part.

So the FEP II worked pretty well and either four or five were finally built. This occupied me and a subset of the design group for several years. But in the fall of 1974, after the first FEP II was up and running, TI had to tighten its financial belt. A dictum was issued to cut back people based on years of service.

I couldn't believe it when I was told to lay Mike Bunyard off. He literally drew the black bean, as the personnel cut was based strictly on seniority and not on worth. So one of the premier digital designers at TI was let go. It truly broke my heart. What a way to reward a key engineer.

There was another incident in 1974 that I didn't observe, since it took place in the TI Board Room during a high-level meeting, but two eyewitnesses have verified it. It began when an outside "Consumer Expert" was hired into TI to develop the digital wristwatch business. The secret project was code-named "Module One," and after some period of time they had developed and manufactured 50,000 digital watch modules and had ordered the same number of stylishly-designed watchcases from Benrus.

The first production-model watch was taken into the Board Room meeting for the presentation to Mark Shepherd. Mark took one look at the new watch and threw it the length of the room and against the far wall as everyone ducked. "That looks like a piece of *shit* and TI's not going to put their name on it!" declared Mark in his usual roar. Neither of my sources mentioned a reply by the Consumer Expert.

Remarkably enough, the story had a happy financial ending when Charley Clough, Semiconductor Marketing Manager, very cleverly sold the watch modules and cases to an established watch manufacturer. Charley was a hero for turning a write-off into a $2.5 million sale. Good boy, Charley! By the way, what ever happened to the Consumer Expert?

The scientists in the Central Research Labs had been experimenting with a new way to clean the surface of the silicon wafers after the photoresist process was finished, and had been encouraging us to build a production version of their "descum" machine for the FEP II. It would save a step of nasty "wet processing" that was done off-line in fume-hoods with acids. This new process could be done in-line, so to speak, as the wafer moved along the FEP II spine, and it didn't involve any acids or liquids of any kind. Great idea except for one slight drawback—it used a high concentration of ozone to clean the wafer surface. At these levels, ozone was highly poisonous and corrosive to the lungs. According to the safety books, if you got a big whiff of concentrated ozone and didn't die, you'll probably wish you had. The CRL folks didn't bat an eye in trying to convince us that it could be done safely if we just put our minds to it. So Tony Adams got the job, assisted by Claude Head on the software. I lent my management expertise from a safe distance.

That machine had every safety feature known to man. There was no way the ozone could accidentally get out of the reaction chamber. But, of course, Tony managed to get a pretty good snort of it and spent a couple of days in the hospital. More safety features were added.

The machine, which was ultimately a failure for various reasons, was cranky beyond belief. It was especially aggravating in its ability to jam up a silicon wafer that was supposed to be moving through it. If the wafer hung up, correcting the problem involved purging the ozone from of the chamber, and, after verifying that it was safe, removing a side panel and prodding the recalcitrant wafer with a stick to move it on its way. Then, putting it all back together again and refilling the chamber with ozone. What a drag.

Claude Head almost made the machine work by some clever software. When a wafer would stick on the air track inside the machine, he had the software operate, machine-gun fashion, every solenoid valve that wasn't being used. The valves were bolted to the baseplate of the machine and nine times out of ten, the extra vibration would jar the wafer loose. But nine times out of ten wasn't near good enough and it was scrapped, with few tears shed on our side of the TI campus. What's wrong with a little nasty wet processing anyway?

Looking back over one of my few surviving office calendars I find the not-very-cryptic notation of *"Ride!"* in a Tuesday square. This meant to bring my Hodaka *Super Rat* motorcycle on the trailer to the office and prepare to do a little dirt-digging after work. This was about the time of my peak outside-TI interest of off-road motorcycling. By 1974 there was a cadre of TI guys that I worked around who liked to meet after work or on weekends for a few hours of wild and crazy dirt motorcycle riding. I remember J.L. "Jake" Brand, Bobby Hetherington, Lowell Barton, Mark Kluge, Dave Willey, Bob Meadows, and Clyde Golightly.

This interest probably came about because I never really got enough motorcycling when I rode my yellow-house-painted Royal Enfield in our Houston days. Off-road motorcycling, or "dirt biking" was a good way to get some fresh air and was a lot safer than riding on the streets. We had some neat places to ride, such as my mother's farm near Decatur and in our East Texas woods in addition to several close-in ones.

So I started out in 1970 by buying an abused Honda *Trail 90* from a neighborhood teenager. The price was right and nobody, not even Mary Ruth, could complain that it looked like a motorcycle. It looked more like a girl's bicycle after a hormone shot. It had a pressed-steel step-through frame and an automatic clutch. It was easy to convince Mary Ruth that the whole family would enjoy it. First, you get your nose in the tent...

That led to a second off-road machine, a Hodaka *Ace 100*. Much more sporting, as it was a kind of poor-man's motocross machine. By this time, our son David was nine years old, and to my mind plenty old to ride a motorcycle. I began looking for a suitable mount for him.

Some scouting around unearthed a Suzuki street bike, not much bigger than a kid's bicycle, with an 80cc two-stroke engine. The fact that this bike was in kit form and consisted of a stripped-down frame and a large box, labeled "Frozen Chickens," full of parts made it even better. David needed to learn how machinery was put together. And besides, it was fifty bucks. How can you go wrong?

And we had a good time building up David's "new" motorcycle. Various unneeded parts, like heavy steel fenders were tossed and replaced with lightweight plastic ones. The headlight and speedometer were just tossed. We put a racing-style numberplate on the front to make it look cool. And it did look kind of cool. Well, for a hacked-up red Suzuki street bike it looked cool.

Our first lesson in motorcycling took place in our church's parking lot on a Saturday afternoon. On this day, David learned to drive off from a stop using the clutch and throttle, just like a stick-shift car. He only killed the engine about fifty times before he got the hang of it. He also learned how to operate the kick-starter.

David put a lot of miles on his off-road Suzuki and we had a lot of fun together. However, he does remember a riding adventure that began when we arrived together at the bottom of a deep dirt bowl in a wooded area off Northwest Highway, not far from our house. It was a large, wild, undeveloped area in the middle of the city and a fine place to ride. It even had a creek to fall into.

On this occasion, we were riding together and had momentarily stopped at the bottom of the dirt bowl. We gunned our bikes and headed up for the rim. As I topped the brow, I glanced over and saw David stall out and lose control just short of the top. He and the bike flipped over and rolled and tumbled back to the bottom in a manner that I thought exceedingly humorous. I wheeled around and went back down into the bowl to help David untangle himself from his bike. As I got off and went over to see about him, I noticed that his mouth was full of dirt. For some reason this led to a laughing fit on my part that he hasn't forgiven me for yet. But believe me, it was funny.

The Suzuki had one unfortunate invention called a "rotary transmission." This neat and deadly feature allowed a street rider to approach a stop in fourth gear, which was the top gear for the bike, and then merely push down on the foot gear-shift lever to go through neutral and back into low gear, ready to start up again. However, as I found out in the woods in East Texas, it was easy to lose track of what gear you were in as you tooled down the piney trails. If you pushed the shifter down once too often, it put the transmission back into low gear, usually at fifty per. The immediate effect of this was to lock up the rear wheel, making the bike exceedingly hard to control. David and I both began keeping track of the gears after we each discovered this flaw the hard way.

But David grew out of the Suzuki, thank heavens, and we sold it and I bought a Hodaka *Super Rat*. David graduated up to my Hodaka *Ace* and we had almost matching motorcycles to ride. Of course, I'd put a reed valve and a trick pipe and plastic fenders and a lighter gas tank, etc. on the *Rat* so he couldn't catch me.

But I had some great times riding with the TI gang. How can I forget Lowell Barton boogeying down the trail through the woods near the Trinity River, having a ball as he zigzagged back and forth from one side of the trail to the other? And then one of his handgrips came off and he rocketed straight into the woods and hit a tree. We gave him a round of applause and a 9.2 for style.

Or the time I was following J.L. Brand and watched as he went over the edge of what looked like an impossibly steep cliff? I decided that if J.L. can do it, so can I, and kicked the transmission into second gear in preparation. My plan was to push off over the edge into what appeared to be the nothingness of a free-fall, and then let out the clutch with my left hand and allow the engine braking to slow my descent.

I rolled over the edge and began the free fall and told my left hand to let out the clutch. My left hand replied it wasn't going to turn loose of anything, thank you, and would continue to maintain a death grip on the handlebar until conditions improved. And it didn't let out the clutch and I was going about sixty at the bottom. I told J.L. I did that on purpose.

And then there was the time Mark Kluge decided to quit dirt riding. I saw him when he made the decision—he was midair in a monster endo on his Honda 125 at the time. It took several of us to extract Mark and his motorcycle from the bushes.

Of course, I shouldn't talk about Mark since I performed a similar maneuver a year earlier. I had briefly killed my engine in the splash of water in a creek crossing while chasing Bobby Hetherington and was trying hard to catch up. Then my foot slipped off the wet foot peg and my Hodaka sailed off the trail and hit a large bush head-on. It's called a "Flying W" as you gracefully rotate your body over the handlebars and into the shrubbery. I remember seeing the sky go by just before I hit.

The year of 1974 drew to a close as the FEP II project wound down. A few more were being built but they didn't need much help from us, so we began looking around for other engineering worlds to conquer. Or at least engage in combat. Jack Brixey of the TI Photomask department supplied the next technical challenge.

Jack had admired the FEPs and wondered why we clever people couldn't build one like it to process photomasks? Photomasks were the "negatives" that were used to print the patterns on silicon wafers to make computer chips. Instead of being a round wafer of silicon, a photomask was a square piece of glass. The processes were different but similar. Similar enough, it looked to me, that we could use the skills we developed on the FEP II project and do the same for photomasks. In mid-1975, we got the go-ahead to design and build the Integrated Mask Factory, or IMF.

Based on the FEP and FEP II systems, the IMF systems were to have long spines hung with process machines. We knew how to do spines, so mainly we just had to invent a collection of new process machines. And they were a tough bunch of machines, technically speaking.

We started designing and building the prototypes for the first of several systems scheduled for the Integrated Mask Factory. The first system was the Mask Manufacturing System One, a.k.a. the MMS I. This would be for the smallest masks being produced, which were square glass three inches on a side and coated with a thin layer of chromium.

An unscheduled diversion occurred in 1975 when I was contacted by some person in TI as being the only person they could find who had actually seen a point-contact transistor. NASA had sent a request, as a result of a contract with TI's Semiconductor Research and Engineering Laboratories, that we build four original-model point-contact transistors from single-crystal germanium that had been grown in "Sky Lab" as it was orbiting the earth.

I accepted the challenge since nobody thought we could do it anyway, and I knew there was one more person at TI that understood point-contact transistors better than anyone in the world. I promptly looked up D. D. "Mac" McBride's number in the TI phone book. Mac, as you may recall, was the last person at TI to actually build point-contact transistors on the production line in 1954, and he was still at TI.

"Mac," I said, "I've got a little project for you. If I supplied you with germanium dice, could you build four point contacts?"

"Sure," he said. "I've got enough parts in my toolbox at home to build a hundred. Bring me the germanium chips and I'll do it. Of course, you'll have to rig some way to form them since the forming machines went to the junkyard years ago."

Oops. I had forgotten a small but critical detail. There was the minor problem of "forming" the transistors. After the transistor parts were assembled mechanically, the two "points" of the point-contact transistor had to be welded into the surface of the germanium to form the p-n junctions that allowed it to have transistor action. A slight oversight on my part.

I took the four tiny germanium squares to Mac and then headed to the TI library for some really old and dusty transistor textbooks. I found one that had a few paragraphs on forming point-contact transistors and checked it out.

While I accumulated and wired up various borrowed power supplies and a transistor "curve-tracer" oscilloscope, McBride built four fine-looking point-contact transistors. They looked fine but they didn't work because they hadn't been "formed." I hoped that I could overcome that lack with my rat's nest of equipment and the too-brief description in the textbook.

After an hour of gingerly playing with the voltages and currents of the power supplies, I formed the first and only point-contact transistors whose germanium had been grown in outer space. They were really lousy transistors, even for point-contacts, but they were transistors. I could have used a lot more practice on my jury-rigged "forming machine." But I still have the Polaroid snapshot of the curve-tracer screen showing "transistor action."

The thank-you memo that Mac and I got from John Yue, the TI coordinator, said that the transistors would be put in a museum somewhere as demonstration devices. "Project Nostalgia" was a success because McBride hadn't cleaned out his toolbox in twenty-one years.

  This year also generated one of my favorite stories about a TI engineer. The nameless engineer, and I didn't know who it was, had flown to Houston to do some work in the Front End and, as usual, had rented a car and driven to the TI plant in Sugarland. He parked and prepared to get out and found he couldn't figure out how to release his seat belt. It was a new model car and all the usual tricks were to no avail. He gave up on the belt release and tried to slide down and out from under it, which he couldn't do. Then, he tried to slither out the top. Impossible. But finally, about to be late for an important meeting, he took out his pocketknife and cut the seat belt loose.

  I liked this story a lot since I could imagine myself doing the same thing. At least a dozen years later I was relating it to a group of engineers, and after I had finished the story and the laughter had died down, my friend Bob Falt spoke up, "That was me." Then I had to ask the question that had been bugging me all these years.

  "Bob, what did the rental car company do when you brought it back?"

  "Beats me," he shrugged. "I didn't bother mentioning it."

  For our group, the year of 1975 was spent working a little on the fading FEP II equipment, and working a lot on the new MMS I. The MMS I process machines for the Photomask department were beginning to take shape by year-end.

  In the early part of 1976 there was a major reorganization in ASD, the Automation Systems Department. Clyde Golightly, my leader for the last two and half years, had been promoted from the boss of "Mechanical Systems Development" to head up the whole thing, now called "ASD Engineering." My Section's name was changed from "Front End Systems" to "FEP and Mask Systems" to reflect my newest customer, and we now reported to the "Front End Systems Branch" under Jim Moreland.

  I also lost Tony Adams, my lead M.E. I could no longer hide his light under my bushel and he was promoted to head up the

Section "Conventional Assembly" under John Stewart's "Assembly Systems Branch." Tony kept right on going up to VP.

But working for Jim Moreland was not a problem. Jim was a really nice guy, who managed without overwhelming, and was supportive of new ideas and well-liked by all the troops. It was the beginning of a long and pleasant association.

The rest of 1976 found us mostly finishing the design and completing the prototype process modules for the MMS I line of the Integrated Mask Factory. But there was still a little FEP and FEP II equipment, and of course, a few more RCCAs. Whatever. Just so I could design and build equipment. But part way through the year I had an interesting interruption. I went to Corpus Christi and worked on a team with Richard Colley, TI architect, on the design of the new FEP II production line in Lubbock, Texas.

Richard Colley was one of the architectural trio of Ford, Colley and Tamminga Partners who designed the semi-exotic Semiconductor Building in Dallas, hyperbolic roofs and all. Colley was a good friend of both Mark Shepherd and Jim Nygaard, and fit the requirements of a TI architect perfectly. He had no qualms about slugging it out with Mark toe to toe, and vice versa. Both Colley and Nygaard went to Texas A&M and shared that mystic bond in addition to their disregard for authority and convention. Colley was stimulating to work with, to say the least. He would tell you that your ideas were full of shit without batting an eye. And what I found out rather quickly, he was usually right. Not always, but usually.

Mr. Colley would sit at the conference table with a roll of onion-skin drawing paper unrolled and overlaid on the Lubbock building plans, and holding a drawing pencil at least four inches from the point, proceed to sketch an absolutely beautiful idea of how things might be designed. Although I didn't always agree with him on the technical details and reasons, he had immense talent and was the crankiest old bastard I ever worked with.

It was an entertaining and instructive two weeks. It varied from the highs of Mr. Colley telling me, "Well, that *might* work," which was his highest compliment, to the lows of rolling his eyes in disbelief and saying, "Goddammit Ed, I thought you were an engineer..."

But most of 1976 was spent building and installing the first spine of equipment for the Integrated Mask Factory, the MMS I. Designing and making the first of anything new work properly in production was a large and headache-producing effort. And the MMS I was no exception. But by the end of the year it was running to specs, or as a memo related, it was achieving "superior quality levels." Except for the photomask contact printer. That damn thing never did work right. There's always one.

During the installation of the MMS I in its special cleanroom, I was standing in a doorway talking with Jack Brixey, Chief Engineer of the Photomask Department, as my guys tinkered with the equipment and a mechanic from TI Facilities worked on some wiring nearby. *Boom!* was followed by *Whooooooosh!* It brought our conversation to an instantaneous halt. Then we were trampled by a half-a-dozen people exiting the area in an unsportsmanlike manner.

The CO_2 underfloor fire extinguisher system had gone off. Actually, it hadn't "gone off," it was set off by the dumb-ass mechanic working on the wiring. We found later that he was moving the CO_2 system firing box to another wall. For fail-safe operation, all the firing switches were wired in series through normally closed contacts. An accidentally broken wire would fire the system instead of preventing a person from firing it. So the mechanic disconnected the wiring to move it and fired the system. As I recall, it dumped eight big bottles of CO_2 under the floor in about five seconds. It sounded and looked like the end of the world, especially when you had your heads under the raised floor like my guys did.

By the end of 1976, work had begun in earnest on the successor to the FEP II. The three-inch wafer was already becoming a thing of the past and to keep up with the industry, we needed machines that would process the new 100mm, or four-inch wafers. In 1975, my group, at the request of Clyde Golightly and Jim Nygaard, had begun to look at some options under the code name of the FEP-X. But it was not to be. By this time the FEP concept had taken on a life of its own and was deep in the politics at TI. So the development of the newest version of the FEP went to a group with more political clout. It was named the Advanced Front End Project, or AFEP, and we weren't part of it.

Chapter Eleven

My group spent a lot of 1977 on the MMS III photomask system for 4- and 5-inch masks. The MMS I was running and we had moved on to the next phase of the Integrated Mask Factory. By dint of clever engineering, we had combined the proposed 4-inch MMS II and the 5-inch MMS III into the 4- and 5 inch MMS III. It would run either size plate with its cunning air track. But maybe it was too clever by half or more because this misbegotten system turned into The Project That Would Not Die. I still have nightmares about it.

During the year we also sent an FEP II to Freising, Germany. Wayne Sullivan, our secret weapon, went with it and got it running properly, thank you. Wayne probably learned to speak German on the way over.

This was also the year we got involved with the projection wafer printing equipment which was sweeping the industry. The design and construction of such optical monsters was out of our technical capabilities. Way out. So thankfully TI bought them from outside vendors. One we liked was built by the Perkin-Elmer Corporation, and was so damn complicated nobody except Jerry Merryman, TI optical-and-everything-else whiz, could understand how it worked. But we were able to tie into it and interface it to the FEP II and the AFEP lines. Jerry explained to me how it worked several times. I still don't understand the scanning part of it, but I can assure you that I could call Jerry tomorrow and he'd explain it to me again and I still wouldn't understand it.

Chester Nimitz Jr. of the TI Houston fame now ran the Perkin-Elmer Corporation. He left TI shortly after his Houston management stint and went to Perkin-Elmer, a small laboratory equipment manufacturer. Chester took them by the scruff of the neck on a new and fruitful path and they had grown like crazy. By 1977 they were building some of the world's most sophisticated optical equipment. When visiting their plant in Connecticut on business, I would tell the engineers stories about the days when Chester ran the Houston plant. I'm not sure they believed me, since his persona was quite different at P-E. They just hadn't had the opportunity to see him operate close-up like I had. He knew how to run a company.

I left the South Building to go home one fine August afternoon, hopped in my good old Plymouth Barracuda, and was met with the slosh of water. The inside body pan was full of water. And how could this be? A bright and rainless day, yet my car was full of water? And with the windows up and both doors locked to boot? As Tony Adams, limey engineer, would say, "Not bloody likely!" Not even the most determined practical joker would do something like this. I bailed out what water I could with my bare hands and drove home. My shop vacuum took care of the rest, and I left both doors open and fans blowing all night to dry it out before mildew set in.

The next day I walked around the parking lot like Sherlock Holmes but without the funny hat, puzzling over what could have happened to my car. Finally I got a clue. The five-sided valve stem atop a nearby fireplug was freshly scarred. Someone had recently turned on the fireplug. *Fireplugs* are full of high-pressure water. *My car* was full of water. Aha! There might be a connection!

I returned to my desk and called TI Security, who were in charge of such things as fireplugs and nonchalantly asked, "Have you-all been doing something with the fireplugs behind the South Building?"

"Well, yeah, we tested them yesterday. Once a year we have to test them for full flow. Why do you want to know?" And I told him. He replied, "But that's impossible! We park our pickup truck in front of the water so that it doesn't shoot up against any cars in the parking lot."

"Oh, yeah?" I replied authoritatively.

He gave in quickly and eventually TI paid for a spray can of mildew killer and air freshener. It was all I could think of to charge them for.

But to even out getting my car hosed by the fireplug, 1977 brought me the unexpected pleasure of being chosen to join the Semiconductor Division Patent Committee. This committee of about a dozen people in a variety of technical fields met monthly and presented and discussed the incoming patent disclosures for our Division. A "patent disclosure" is the document that is filed with the TI Patent Department by any employee with an original idea that they think could be patented and would be of use and profit for TI. The committee generally handled ten or twenty possible patents each month, with each of us being assigned one or several that were in our field of expertise to discuss and make recommendations for ranking in the next meeting.

It was our duty to analyze each submission, understand it, and talk with the incipient inventor if necessary. Then we scored the idea in such a way as to be able to make a rational group decision as to its degree of goodness. It cost TI a lot of money to carry an idea through to a patent, and there was an ongoing shortage of attorneys, so it behooved us to choose carefully. But not so carefully that we turned down a sleeper that might be a big-time winner.

Most of the patent disclosures were easy to rate. Some of them were just plain dumb or were not patentable, and a few were so complex and esoteric that none of us could decide if it was a good idea or not. The final result of our discussion of a disclosure was a ranking number. If the patent attorneys all caught up with their backlogs next month, even a lower-ranked idea might be filed with the US Patent Office. But with the usual pile of patent work, it was the other way around—some pretty good ideas never got off the table for lack of legal horsepower.

So after careful study of one disclosure assigned to me, it appeared to be a loser and I so scored it. It was a clever tool for use in manufacturing but it really had no monetary value as a patent. TI wasn't in the business of selling clever tooling. The head of our SC Patent Committee asked me a second time, "Are you sure you want to scrap this one?"

"Yes," I said, "Even though J. Fred Bucy turned it in, it doesn't meet our requirements."

"Well," said our chairman, "I'm going to override your decision. Since J. Fred is the president of our company, I think he should get a patent on this." All the other members nodded in agreement. It just goes to show, numbers aren't everything.

This brings to mind the 1970 patent that I shared with Boyd Cornelison and Fred Romberg on the hourglass lunar gravity meter. This, shall we say, unique idea was first rejected as a TI patent by a committee much like the one I was on. I have a copy of the rejection notice in my files. But when someone higher up noticed Boyd's name on the disclosure, the committee's decision was reversed. The patent folks had a standing order from Mark Shepherd to patent anything Boyd Cornelison turned in. Just because the patent committee thought it was a rotten idea didn't make any difference.

The story goes that the TI attorneys had once failed to file for a patent on an idea submitted by Boyd and it turned out to be an industry-wide Goodie with a capital G. Shepherd was pissed and told the lawyers that they didn't need to puzzle over Boyd's ideas any longer—just patent anything and everything he submitted. He probably said it a little cruder than that, but it was a valid dictum. Numbers shouldn't be everything, especially where Boyd was concerned.

I stayed on the Semiconductor Patent Committee until 1984 and thoroughly enjoyed it. Each new disclosure that was assigned to me was like getting a birthday present—it could be anything. Sometimes ordinary, sometimes extraordinary, but never dull. Someone at TI had entrusted me with the baby that they had given birth to and it was an exciting time for them. They had put the most important thing in the world into this disclosure—their ideas—and it was sent to me to decide if it was to live or die. The form letter we sent to those hopeful inventors that "the decision of the Patent Committee is that there is not sufficient merit..." was always a blow. Lord knows I'd received enough of them myself.

But this was balanced by the letters stating, "...the Patent Committee has decided to pursue your Disclosure Number..." Now *that's* exciting!

During 1977, the Photomask Department consulted with us about a growing problem. Their large "library" of master photomasks and reticles was growing by leaps and bounds, and it was purely a hand-and-foot-powered "store stuff in boxes and get stuff out of boxes" operation. It was getting out of "hand," to be cute about it. So with the encouragement of Photomask and a little under-the-table money, my group set about, in our spare time, to gin up some ideas for an automated mask and reticle storage facility.

This was a fun kind of project, since we could brainstorm to our heart's content and were not expected to produce any serious hardware. Just kind of noodle around with ideas and see what happened. It was like having a little dessert after being good engineers and eating our spinach. And Lord knows there were days when our diet was all spinach. So we all joined into this low-level invention process and stirred the cooking pot of ideas, throwing in bits and pieces of all types.

We built a breadboard of storage bins with a tracked robotic handler that ran up and down the aisle, picking and replacing the delicate and fragile glass plates. We had a ball doing it, and it was a pleasant diversion from our usual grind. But, alas, it was not to be. The calculated savings would not justify the cost, and the Photomask Department did not pick up the project. We even had a swell name for it, "The Crane Brain." Unless, of course, that's what scared them off.

Working with Jack Brixey of the Photomask Department was a pleasure. Never excited, always calm, he dealt with us directly and fairly. His chief assistant scientist, Dr. Bill Almond, had a different temperament. Bill always had a million ideas and was excited about them all. We couldn't do jobs fast enough or think big enough for Bill. He stimulated our thinking beyond the breaking point on occasions, but he was a good balance for the group.

Jack Brixey, long-time TIer at this time, was also very active in the Boy Scouts. His one great invention that he never followed up on, was the kerosene-powered transparency, or "foil" projector. TIers were so accustomed to using foils for their presentations that Jack wanted a non-electric model that he could use in the woods while camping out to explain things to the Scouts.

I was told, on pretty shaky authority, that Texas Instruments was the second largest user of the 3M Company's transparency material, with the US Government being number one. To this day it's second nature for me to put the foil down the proper way on the projector. You can always tell someone who's never worked at TI during any kind of meeting with a transparency projector—there are three wrong ways and one right way to put your page down for projecting.

1977 slid quietly into 1978 as the Mask Manufacturing System Number Three, a.k.a. MMS III, droned on. We were well into the design of ten process modules for the system and I had vowed that we would finish the goddamn thing before the end of the goddamn year of 1978. Keeping money flowing from the Photomask Department to support my group was an ever-increasing part of my duties as manager. Fortunately, Jack Brixey and Bill Almond, pillars of the Photomask Engineering Department, were on my side and wanted to see the MMS III get built and into production, even if it killed me.

But, as usual, there were some entertaining sideline activities that served to keep me from being pushed over the edge by the MMS III. One developing interest was the effect of a building's floor vibration on the operation of the projection wafer printers. I began working with the manufacturing people on their vibration problems and soon discovered that there was no one I could find at TI that knew much about it. Vibration looked to me like a good field to get into. Maybe our group could turn a few bucks doing vibration reduction or something. The one thing I knew for sure was that things were going to get worse before they got better on the manufacturing floors—each year the patterns on the silicon wafers got smaller and more difficult to print and there wasn't a coordinated effort on the vibration problems.

Another sideline activity that I couldn't resist was a "Name the Newsletter" contest sponsored by Ron Jones, head of the recently-announced Dallas MOS III "front end." This front end, which was being built in the South Building, was coming together fast under Ron's firm hand. People were hired and equipment was being installed. It was time for a newsletter to give them a sense of unity and to keep the employees informed of the rapidly moving events of D-MOS III. If there's anything I'm a sucker for, it's a "Name the Something" contest. And besides, the prize was a new TI digital wristwatch. Lord knows, I needed a new wristwatch. So I entered the name "Dallas MOS !!! Newsletter" and won. It was truly a proud day for me and I still have the watch. It doesn't run anymore, but I still have it.

Later, the installation of the equipment into D-MOS III began to slow down, and then it stopped. The hiring of people dwindled to nothing, and rumors began to be whispered about. The next newsletter came out grimly headed, "Dallas MOS ??? Newsletter." And then the whole project went down the tube. Sad. But it was not for lack of leadership by Ron Jones or a good newsletter name.

Ron ended up temporarily as building manager for the South Building and, sure enough, one of the first things he did was to instigate another contest. This one was for a slogan to remind people to keep the building tidy. Hot damn! I could always use another watch. I entered about 50 slogans, such as, *"Keep the South Building clean or I'll kill you!"* and *"Don't drop that shit on the floor!"*

You may not believe this, but I didn't win. The winner was some threadbare cliché like, "This is your South building! Keep it clean!" Ron later told me confidentially that he didn't think it would be fair for me to win two contests in a row. That made me feel better, but I still think he could have picked a more imaginative one, like "Please don't litter!" Some days I just waste my talent...

I did have one creative winner during the year, which certainly didn't include the MMS III. My long-time friend and super-duper electrical engineer, Lee Blanton, was to celebrate the 20th anniversary of his employment with Texas Instruments. What a fortuitous occasion for TI! And what a fortuitous circumstance for me, for whenever Lee worked for me I got really good reviews and raises. I was determined to give Lee an outstanding gift for his anniversary. Anybody who could last 20 years at TI, especially working as long and hard as Lee had, deserved it.

One of Lee's pre-TI employers was the US Navy. Lee once told me an interesting story about being on shipboard with Dr. Van Allen when he was studying the ionosphere and first detected the ionized band now known as the Van Allen Belt. Lee was just a sailor and not directly involved with the launch of the high-altitude rockets for the experiments, but he thought it was a neat experience to be there when it was happening. I suddenly figured out what to do for Lee on his anniversary.

I called Clif Penn, ever-useful scientist in the Central Research Labs, and left him a message. "Do you know if Dr. Van Allen of the Van Allen Belt fame is still alive, and if so, where?" Clif's phone call back was taken by Tom Bimer, one of my engineers, and the note to me read, "Dr. Van Allen is alive—head of Physics at University of Iowa in Iowa City." Bingo! The phone number was easy to get and I called Dr. Van Allen.

After explaining to Dr. Van Allen who I was and what I was trying to do, he graciously agreed to go along with the gag and would get a letter out to me posthaste. And he did. A very pleasant gentleman indeed, to go to this trouble for an unknown caller with a really strange story. I got a large envelope within a week from Dr. J. A. Van Allen, The University of Iowa, Physics/Astronomy Department. In it was a letter addressed to Lee Blanton.

At the badge party for Lee's 20th, I handed Lee the letter from Dr. Van Allen and a gift-wrapped box from J.C. Penney. I asked Lee to read the letter out loud to the audience. It went something like this, "Dear Lee Blanton, Congratulations on your 20th anniversary with Texas Instruments. As we worked together years ago on shipboard while I was making my discoveries, I'd like to personally thank you for that work and to honor you for your service at Texas Instruments with your own personal Van Allen Belt." It was brown leather, size 34. Lee always was slim.

Towards the end of 1978 there was a flurry of memos between Jim Nygaard and J. Fred Bucy, the TI president, about training the operators on the Perkin-Elmer projection printers. (Actually, Jim sent "memos" and Fred sent "BucyGrams." Ask anyone who worked for Fred.) The P-Es were difficult machines to run and required considerable operator training and this training meant they had to take a Perkin-Elmer out of production. Even worse, the minimal training given to the operators resulted in a lot of faulty material for the first weeks or months on the production line. Jim and Fred kicked around the idea of building some sort of training machine, and Jim included me in the conversation. Nothing directly came from the conversation, but the idea remained lodged in a corner of my brain. What a neat project that would be! One of these days...

The year of 1978 ended a little like it had begun—working on the MMS III. I had failed in my oath to get it done or else. So, it was "or else," whatever that was. But "or else" didn't mean that the Photomask Department was going to cancel the project, stop paying my group, and then stick the MMS III up our collective asses. *That* was the good news. Ever onward! Excelsior!

So 1979 continued the world's longest project, with no end officially in sight. A few prototype MMS III process modules had been completed, others were still under construction. Alas! Would it never end? Stay tuned.

My interest in the influences of floor vibration on the Perkin-Elmer printers was growing. The problem was actually pretty simple. The projection printers built by Perkin-Elmer were optical marvels as I have already mentioned. They could do optical tricks that defied Physics 100. They could, that is, if you didn't jiggle them when they were scanning the microscopic images onto the silicon wafers. And guess what? TI's semiconductor manufacturing areas were almost all on the upper floors, with post-stressed concrete floors supported by columns. This construction, suitable enough for the earlier printing technologies, now made wafer printing similar to taking pictures while jumping on a trampoline.

I acquired some excellent vibration measuring equipment, namely seismic vibration pickups and a Hewlett-Packard Fast Fourier Transform "FFT" Spectrum Analyzer. My first test was to make a vibration survey of the South Building floors.

The basement floor, which was a thick concrete slab poured on top of prepared soil, showed a vibration, or vertical movement, of about one microinch, a millionth of an inch, which is rock steady. The first floor, or ground floor, supported by a forest of closely spaced heavy columns showed movements of about ten times as much, in the order of ten microinches. The manufacturing floor, which for ease of equipment placement had widely-spaced columns and an expanse of unsupported floor, was ten times wigglier yet, at about a hundred microinches. And this was what the Perkin-Elmer printers were supposed to sit on and print tiny lines on silicon. Not likely.

This led to a nice business for our group. A modest amount of our engineering effort developed a vibration isolation cradle to go between the Perkin-Elmer and the manufacturing floor. It used air-filled cushions and worked like a champ. We began selling them a few at a time in the latter part of 1979, and as word got out that they worked, more orders came in.

In addition to selling the cradles, I would, for a nominal fee, do a simple vibration survey of your manufacturing area and tell you where to stick your printers. It didn't take the production people long to figure out that they didn't need me for this information. The best you could do was to put the cranky things as near the columns as you could crowd them. The vibration was an order of magnitude lower there than out in the middle of the big concrete drum-head.

So I began developing more sophisticated methods of vibration analysis. Something that looked a lot more scientific than telling them to move their printers over by the columns. A person couldn't make any money doing that. But, actually, I did need a new way to test the floors. What I really needed to measure was the compliance of the floor, or as we say in scientific talk, how bouncy is it?

After a few easy calculations, I designed and built a "thumper" that would stimulate the floor into vibrating and telling me all of its secrets. Nothing could be simpler: thump the floor in a known and repeatable manner, and then see what the floor does. What the floor does, is go "Boing!" when thumped, with the "Boing!" at 12 to 16 Hertz in most of the TI buildings. And if the floor was well-supported, like near a column leading to the ground, the response was more like "Boing!" In the center of a large unsupported span of flooring, it was *"Boing!"* It was easy to thump the floor in a nice grid pattern and then plot contours on a map showing the stiffness of the floor.

The upshot of these baby steps in vibration analysis was that our department set up a full-time vibration analysis person, Stan Pritchard, with good equipment and sicced him on the manufacturing areas. He stayed busy continuously for years, traveling all over the world doing vibration surveys in TI facilities. It was a very useful thing. The problem of vibration grew worse every year as the patterns on the wafers got smaller and more difficult to print.

We, meaning my group of engineers and technicians, found another winner in the Perkin-Elmer printers. In addition to the vibration cradles we'd whipped up, it seemed that they also needed interfaces to the AFEP lines. So, another long-lived product was born. The P-E interfaces, for various models of printers, were built into the early 1980s. At nearly three times the price of the vibration cradle they were a useful addition to the financial well-being of my group.

It was just about one year before that Jim Nygaard and J. Fred Bucy had exchanged thoughts about building a training machine for the Perkin-Elmer production operators. In the meantime, the TI 99/4 Home Computer had come out. First announced to the public in June of 1979, accessories were being added to the product line regularly. It looked like we had an in-house computer that we could use to build a simulator for training P-E operators.

The TI 99/4 Home Computer had graphics capabilities, although by today's standards the capabilities would be considered less than rudimentary. Archaic, possibly. But with the team of Jim

Sims, ace Ph.D. Electrical Engineer, Wayne Sullivan, now upgraded from ace Electrical Technician to ace Electrical Engineer, and John Hayes, ace Electrical Technician and Software Whiz, the problem didn't stand a chance.

In addition to the obvious design of the computer control of "pictures" on the display screen that would look like the operator's microscope view of the alignment process, a mock-up of the Perkin-Elmer operator's station had to be built. The trainee needed to experience the same body, head and eye positions, the same controls, and the same feel as the real thing. We wanted to do as complete a simulation as possible.

We purchased the same joystick assembly as was used in the Perkin-Elmer and mounted it in a mock-up of the Perkin-Elmer operator station. In place of the operator's microscope, a viewing hood was built that positioned the head in the same location as the microscope, and required the trainee to look down at the same angle. But instead of looking through a microscope, the trainee would see the CRT monitor of the 99/4 through the darkened cone, about two feet away, or much the same as the virtual image in a microscope.

I never understood exactly what our engineering team did to the 99/4 inner workings and the modifications made to the plug-in programming modules, but they worked. The graphics system of the little home computer was substantially modified. It could now very nicely emulated the alignment experience of a real Perkin-Elmer optical printer at six or seven percent of the cost.

The trainee would see the two patterns overlaid in the "microscope": the fixed "slice alignment mark" and the movable "photomask alignment mark." By operating the joystick, the movable pattern was aligned over the fixed target and the simulated "Expose" button pushed. The computer would then measure the accuracy of the alignment. After a series of alignments, the game was over and the trainee received a score for the total goodness of alignment accuracy. And sure enough, the next time would be better.

The upshot of this neat system was the ability to train production operators off-line without tying up a $150,000 Perkin-Elmer aligner. Just as important, operators trained on the simulator hit the production line running, so to speak, with a much higher level of skill than previously. Considering the cost of the training system versus the time value of the P-E and the cost of ruined wafers, it was one of the big bargains of the year.

I feel the need to quote from a memo sent to me the 7th of October, 1980, by Perry Denning of the TI Houston facility:

Preliminary results on the A&E [Align and Expose] *Simulator indicate that a substantial reduction in training time is possible thru use of the simulator. The skill developed is that of fine alignment of alignment markers. One inexperienced operator was able to reach the skill level of an experienced operator, that is greater than 3 months, in fifteen hours on the simulator.*

You don't need field sales with customers like that. We built ten systems and the various manufacturing groups put them into action training operators in 1980. They liked them, and we went on to bigger if not better things and forgot about them.

In 1988 I got a phone call from one of the Front End training departments, "Could you supply us with another modified plug-in module for the 99/4?"

"For *what*?" I asked.

"For the Perkin-Elmer training machine," was the reply. I had no idea they were still running. Not only running but they were concerned that we couldn't supply spare parts for them. And we couldn't. Whatever information we might have had on the system was long gone. And the last 99/4As were built in late 1983. Sorry about that. But it was a pleasant surprise that a few of the systems had been quietly running and training operators for lo these many years. Jim Sims, Wayne Sullivan and John Hayes had truly done good. But that was no surprise to me. They *always* done good.

And 1979 was about all used up and the MMS III still wasn't finished. The Project From Hell was like a mirage on the desert—we just thought we could see the end of it. Most of the process machines were finished but a few were still being built. Several turtles passed us along the way.

Almost everyone left for the Christmas holidays but for some reason I stayed and worked. Maybe I thought I should use the holiday time to repent for the MMS III and other engineering sins I had committed. It was nice and peaceful and a good time for repenting and to get loose ends wound up by the year-end.

Whatever I did wasn't exactly monumental since the only thing I actually remember doing was getting the department mail every day or two. As the ranking member of the engineering staff, it was my duty to traipse down to the pipe space and wend my way along the dimly-lit and circuitous path to the mailroom. After a knock on the window and shouting out my cost center over the rumble of the air handlers, the mail person would grab the armload of mail out of our cost center box and load it into the automatic bundling and tying machine. On this day, he dropped the mail into the machine, pushed the start button, and left to do other things.

I watched the Rube Goldberg contraption as its spindly tying arm, trailing a heavy string, swooped around the bundle of mail. As it circled the mail, the arm inadvertently snagged the line cord and jerked it out of the wall plug. Then, as the machine coasted to a stop, it tied the mail bundle up with its own electric cord. It was a clear-cut case of mechanical suicide.

As we plowed full speed ahead into 1980, our group expanded to pick up the additional work with the Perkin Elmer printers. My nemesis, the MMS III, was still with us, but the other paying diversions were increasing and were a welcome relief. But as we expanded to meet the growing internal business, the business within TI suddenly began to slow down. Money, so free a few months before, had now begun to dry up.

The good news was that the Photomask department had invested so much money in the apparently mythical MMS III that they couldn't quit now. But the completion of the MMS III was in sight, or had I said that before? I began beating the bushes for other things to build or services to perform since the MMS III only supported part of my engineering group.

Back in November of 1979, my new group had been transferred from Jim Moreland to Jim Harper. Jim Harper, or Doctor Jim Harper to be exact, was a fun guy to work for. His last project had been on the TI digital watch program and under his direction the group had developed a unique monolithic blob that contained all the working parts of the watch. It was known locally as the "Funny-looking Integrated Circuit," or "Funny-looking IC" for short. He, then, was known as the "Funny-looking IC Manager." But he was a good guy to work for. His Ph.D. in metallurgy was a non-trivial accomplishment and as I've said before, it's more fun to work for a smart person—and he was.

And then, as they say, a miracle occurred. The engineering groups under Jim Harper were split up and reorganized, and according to Jim's memo, Harry Waugh "assumed Ed's role as IMF Project Manager." What Jim Harper's memo meant was that Harry "assumed Ed's position" with the MMS III. He chose his words carefully. In either case, it meant that I was no longer responsible for the MMS III. Gee! And after I did all the hard work!

It was a blessing for me to turn it over to Harry even if I did get it back later. He was a good manager and had done a large amount of the engineering. The MMS III actually was on the downhill side, after all those years, and Harry was a good choice to finish it up professionally. The year of 1980 ended with the MMS III mostly running and partially evaluated, complete with on-line plasma etch chambers. The Project That Would Not Die finally, *finally*, stirred and came to life. It wasn't quite over yet, but thanks, Harry.

While searching my files for information on this time period at TI, I ran across a copy of my memo, dated 19 September 1980, answering a Photomask request that I explain why I was years late and a gazillion dollars over budget on the MMS III. My response was really, really well done.

The last information I could find on the MMS III was a budgetary input for engineering sustaining through October 1981.

The same memo from Jim Harper that turned over the Photomask project to Harry gave me the title of Engineering Manager, Advanced Front End Equipment. I now had the responsibility for all of the Advanced Front End Project—AFEP—that had been "released"—put into production—including the software. We weren't involved in designing it but now we had to keep it running. Whatever. It all paid the same, and it paid, which was the main thing.

So part of my group spent the year upgrading the AFEP equipment, both hardware and software. Field service kits of upgrade parts, field service bulletins, and software patches were cranked out by my group for several years. We supported the hell out of the AFEP equipment.

But the year of 1980 started out as a lean time for my group, financially speaking, and my hardest job was finding paying work. Somehow my branch had grown to 14 salaried, or "exempt" engineers and 11 hourly-paid, or "non-exempt" technicians and draftsmen. Why do you always end up with a really large and talented technical group just as your customers run out of money? It was just like being a small company in a big city looking for business during a depression. And we had a lot of mouths to feed.

The vibration business was still making money, and we continued to sell the air-lift vibration cradles for the Perkin-Elmer printers. And of course, our new Perkin-Elmer Simulators were being built, but I spent as much time as I could in the customer's operations looking for opportunities. I walked back and forth between the South Building and the Semiconductor Building a lot.

As I walked between the buildings I frequently saw my favorite pair of bumper stickers. They were on a pickup truck that parked in the TI lot on my path. The "Hers" sticker said, *Honk if you love Jesus,* while the "His" was, *You can have my gun when you pry my cold dead fingers off the trigger.* Or, I suppose it could have been the other way around but I don't think so. Just another happy, compatible Texas couple.

One day, wandering in an integrated circuit manufacturing area, I noticed a strange attachment perched atop a Perkin-Elmer printer. I asked the line foreman and was told that it was "A Perkin-Elmer Top," which didn't exactly tell me all I wanted to know.

A little investigation elicited the fact that this "Perkin-Elmer Top" was indeed the latest and hottest accessory from P-E to make their printers work better, and TI was interested in buying a bunch of them. Hmm. Didn't look all that complicated. You don't suppose...?

By this time Harry Waugh had the MMS III under control, which was something I was never able to accomplish, so I enlisted his aid. We cooked up a plan—we would build the P-E tops instead of buying them from P-E, thereby making our group a lot of money and saving our collective butts. Our plan had everything: a need, an action, and a result. We thought the result of saving our butts would be outstanding if we could pull it off. The plan was put into action.

On my next trip to the Perkin-Elmer plant in Connecticut, I took the printer project manager aside and asked him the delicate question if P-E would mind if we, TI, copied the P-E tops and built them ourselves? Looking back with a better perspective, I can see why his answer of "Gosh no! Go right ahead! Do you want the **parts list**?" was not as surprising as it might sound.

I was slightly startled by his ready agreement not only to let us build them, but to help us. But TI was a good customer of P-E and had spent literally millions of dollars on their printers. The loss of a few hundred thousand bucks of sales on the tops was nothing compared to the good will and continuing happy association with the deep pockets of TI. In any case, Perkin-Elmer, as usual, was a joy to work with, especially since they were going to let me build one of their products for TI and pocket the profits in my group.

So Harry and I and others caused Perkin-Elmer tops to be built in substantial quantities. It was a nice straightforward product with no high-tech quirks in it. Just sheet metal and a medium-sized sack of electrical parts and some wiring topped off with a nice particle-free paint job. The TI customers liked them too, since they were not only cheaper than P-E's quoted price, but we were right there within striking distance if something went wrong. It was one of our bread-and-butter products well into 1983. We built dozens of them and sold them for what I understand is called "fourteen long" in the gambling world.

During an otherwise uneventful day I got a call from the TI Patent Department. They were in a bind and needed my help. They had a "walk-in" inventor and wanted someone to evaluate his idea. They were all terribly busy but were certain that I could handle it, and they had already signed him up on the usual forms absolving TI from any liability if we should suddenly invent what he was peddling two weeks hence. So they had already done the hard part and if I could just come over for a few minutes...? Those lying bastards. They knew exactly what was going to happen but I didn't, so I went over to help out my old buddies.

I walked over to the Central Research Building and was introduced to the drop-in inventor and also to a TI scientist who I didn't know. The "inventor" was dressed like he was going bowling with the boys but I put this down as the usual penniless genius. The patent people had found an empty conference room for us and the three of us went in and made ourselves comfortable. Our inventor spread out a sheaf of papers and diagrams on the table and handed us a document.

He began to speak to us about his invention. He sure as hell didn't sound like an inventor. He sounded like what he later told us he was, a truck driver. I didn't like his opening words, which were, "This isn't really perpetual motion even if it looks like it at first."

It went downhill from there. His explanation of what made his motor run was a combination of words taken randomly from an advanced physics textbook and a pamphlet describing the operation of an electric motor. In fifteen seconds it was obvious that the visitor was seriously demented. Plus, he was really intense about his invention. Plus, he was big and burly, like a truck driver. It added up to a really poor combination. No wonder the lawyers had called for outside help and were now probably cowering in their locked offices.

The more he talked about his invention, the more animated he became and the closer my mute scientist friend edged toward the door. The demented one finished his meaningless description of the workings of his motor that put out more power than it consumed, and told me in no uncertain terms that this was the greatest invention of the century, if not of all times. If TI didn't see that and buy it from him for a lot of money, he'd goddamn well like to know why. He emphasized the "why" by banging his folder on the tabletop. A quiet click of the door shutting told me I'd lost my only ally and I now stood alone, or sat, actually, facing a really scary person. I began the only counteroffensive I could think of.

"Well, I'm going to tell you a story you won't believe. TI is *so stupid* that it won't even consider outside inventions. And I can tell you, they're going to miss out on a great one here."

"You mean they won't even *consider* it? After all the trouble I went to?"

"That's right. TI is *so stupid* it couldn't see a billion dollars right in front of its face. Can you believe that?" I hoped I had rolled my eyes convincingly.

"Boy! That's *really* stupid! Why the hell do you work here, anyway? I wouldn't work for any company *that* stupid!"

"Well, I have to work here. It's in my contract. But as soon as I can, I'm going to get out of here and find a good company to work for. Why don't you get your stuff and let's get out of here? This is just a waste of your time. You could be working on your next invention or something." I got up and headed purposefully for the door and he followed.

219

We walked to the guard's desk by the outside door, side by side, friends united against a common stupid enemy—Texas Instruments. I smiled and waved to him as he left. My God, I was glad to see him go. Then I went to find my friends in the patent department.

We wound up 1980 by selling just one more RCCA and with Jim Sims building some special test doo-hickeys for the 99/4 computer production line in Lubbock. Never did quite understand what he was building. So the year turned out okay after all. We survived and had a nice line of products and happy customers, if you don't count the Photomask Department.

As we began 1981 we transferred back to Jim Moreland and his familiar and comfortable environment. At the same time we picked up a customer who really liked working with us, and vice-versa. Jim Lewis, boss of the Houston Bipolar 4 integrated circuit production facility, wanted us to support his AFEP equipment and his Perkin-Elmer printers. He couldn't have picked a better or more deserving bunch. We were experts in the upgrades for the AFEP and whizzes in all sorts of things related to the Perkin-Elmers. And, of course, we had one more feature in common—he had money and we needed money. It was a match made in Heaven, er, Houston.

About the time we got cranked up good and settled in, it was time to move. This time, it was the really big move to the Floyd Road Building. The suggestion had been made several years before that all of the semiconductor support groups be moved out of the main TI buildings and clustered in a building of their own. There were four large groups: the Front End Systems (which we were in), the Assembly System (which did the packaging on the integrated circuits) and the Test Equipment Group, whose function was obvious, and the big machine and sheet-metal shops. In addition, we had the usual computer and software group, purchasing, production planning, human resources, accounting and I don't know what else. We could pretty well fill up a big building.

So a building had been built and we were moved into a space on the mezzanine. That was the good news. The bad news was that the building was built as a warehouse, not an engineering and manufacturing facility. It was lacking some of the amenities of a real building, such as an elevator to the mezzanine. But we moved in anyway, and the improvements could and would come later.

John Hayes, our whiz-bang software tech, couldn't climb the stairs and had to be separated from our group. He and his desk were stashed among the department's computer people on the ground floor until an elevator was installed a year later. That kind of inconvenience was countered by the advantage of finally getting everyone that supplied the semiconductor departments with equipment within shouting distance. The bad news was that our customers were out of walking range. More good news was that the automobile parking places were a lot closer.

One feature of the mezzanine that caught my attention immediately was the low resonant frequency of the floor. Since I'd been thumping on TI floors for the past couple of years, I was sensitive to such things. The mezzanine floor in the new Floyd Road building was the springiest floor I'd ever walked on. A few quick thumps and a look at the FFT analyzer showed a resonant frequency below ten Hertz, which easily set a new record for flimsy flooring. And it took a little getting used to. Anyone whose desk was near the center of an unsupported floor section, that is, not near a support column, was subjected to a noticeable oscillation.

The floor problem was aggravated by Cecil Davis. Cecil, now the boss of his own engineering section, resided at the far end of the aisle that I adjoined. Since Cecil never moved at anything less than a 5-mph race-walk, I could feel him coming down the aisle from inside my office. His natural pace was a submultiple of the floor's frequency, and when Cecil headed down the long aisle, he excited the vibration of most of the mezzanine's floor system. One software person actually had to leave the area periodically and sit in his car to keep from getting nauseated. I wish I'd thought of that.

But otherwise our offices on the mezzanine were just fine and we scarcely missed a beat during the move. The AFEP sustaining for Jim Lewis and others was keeping us busy, along with the now-famous Perkin-Elmer tops. The MMS III work was tailing off, finally, but it still kept a few people busy and a few bucks coming in. And of course, we had built a bunch of Perkin-Elmer training simulators that got the year off to a good start.

And to top it all off, Claude Head, my software engineer friend, and I had found a swell barbecue place in Richardson for lunch. There was no cafeteria in the new Floyd Road building, so we skipped out almost daily for BBQ. But one day, as we left the BBQ joint, Claude's good old Buick station wagon wouldn't start. Like any electrical engineer worth his salt, Claude had been fiddling with a worn-out car battery to see just how long he could get by with it. Well, he had fiddled one day too long. His car wouldn't crank at all.

Claude, totally undaunted, got his jumper cables and opened his hood. He looked at the unoccupied cars on either side and made a snap decision—the one on the right. Claude walked over; popped open the hood of the unoccupied car, and connected up his jumpers. While I stood on the curb, mouth agape, Claude got in his car and cranked it up. He removed the jumper cables, slammed the hood of the neighboring car and said, "Thanks!" to it.

If I had tried that, the biggest guy in the world, who had just gotten a traffic ticket and found out his wife was leaving him, would have come out of the BBQ place and pounded me into the ground like a tent stake.

So 1981 finally went wherever used years go. The economy had perked up during the year and the tight days were over. Things were again running smoothly at Mother TI. My group had sold a third of a million bucks worth of products and services and the MMS III was finished. Hallelujah!

Chapter Twelve

A ny year that starts off with a party at work can't be all bad, and 1982 started off with a dandy. It was a badge party for my 30th anniversary of employment at Texas Instruments. And they did it up brown. The goal of badge parties in those days was to embarrass the honoree and they succeeded. I have a videotape of the whole thing and can prove it. They spared no effort..

But, deep down, it really was a nice honor. Not many people had lasted thirty years at TI, even bopping around from department to department like I had. And I had escaped entirely for a year to Textool for a cooling-off period. But by the unfathomable algorithm used by Human Resources, the last day of 1951 was my official employment date, and so the last day of 1981 was my thirtieth anniversary date.

My party was attended by just about everyone in the Automation Systems Department, including and especially machinists from the machine shop who I didn't know. Everybody went to the badge parties for a little R&R and some punch and cookies. It didn't really matter who the party was honoring. These parties had evolved over the years, mostly from the efforts of Jim Nygaard. His idea of a good badge party was to do something really silly, and so over the years, we did a lot of really silly stuff. If you recall my story of the "bull wheel" in the 1960s, you know what I mean.

Conversely, when it was someone else's badge party, I enthusiastically spent money and time on the silliest things I could think of for them. I can recall a few I built and presented over the years:

- A pair of crutches for Harry Waugh, with ball bearing wheels on the tips and equipped with an air-speed indicator. If he broke a leg skiing, he would still be able to walk fast.

- An "Aggie Ring" for Gib Hagler, resident Aggie engineer It was fitted with an internal mercury switch and tiny battery and lit up when he picked his nose.

- And my favorite, a decal for Jim Nygaard to put on his Briggs and Stratton lawnmower engine that read, **FIRING ORDER — 1.** I've still got the spare I made on my mower engine.

And my party was a fine party. The big broadside at the end of the room proclaimed the day to be "Millis Mania" and a screen and projector were set up for showing embarrassing pictures. Mary Ruth and our son David, 19 at the time, were invited to join in the celebration. As I soon found out, Mary Ruth had sided with the enemy and supplied the pictures that started off the show.

After my baby pictures had been shown and the guffawing died down, my "friends" came up to the microphone, one by one, and presented their special gifts. They ranged from really clever, like Mel Creps's "Chainlite Home Saw" (ask me about *that* some day) to kind of obscure, such as the box of ice to sit on the next time I dug a well. Get it? The Klondike well digger?

After I was handed the penultimate gift of a lit stick of "dynamite," Irwin Carroll, MC, invited Floyd Purvis to come up for the final presentation. Floyd was the head Security honcho at TI. He and three other guys took over and proceeded to deck me out in a safety outfit so I could walk across Floyd Road without harm. This took a while, as I had to put it all on. Or, rather, I had to stand still while they put it on me.

The yellow slicker and rain hat were followed by a huge hand-held "STOP" sign, followed by a safety flare (unlit), a flashlight, and a hard hat. The grand finale was a shoulder-supported "sandwich board" sign that said, "Please Stop" on the front and "Stop Dammit!" on the back. This was all because I had sent a memo to Purvis in December complaining about the lack of a safe way to cross Floyd road on foot. I refrained from pointing out that for the same amount of money and effort they spent on my "gifts," TI could have installed a traffic signal at the corner.

This was a typical badge party of that day, and it was a nice social gathering. Rarely was anyone seriously injured by his or her gifts. But the custom began to fade and rules were instituted that limited the size and extent of the parties. If I had stayed around for my fortieth, it would probably have been a nice gift from TI, a new badge, another anniversary pin, and a hearty handshake from my supervisor. It would have been a lot cheaper for TI but not nearly as much fun.

At the encouragement of Jim Moreland, I began pursuing the idea of a clean carrier box system for moving the silicon wafers in production between the processing machines. Management was convinced that he who first automated the movement of wafers in a computer chip factory would rule the world, and that the first step on this slippery slope would be a standardized ultra-clean container.

The container was more than a cute box, as it had to protect the wafers from the particles in the air, even in the cleanroom air, and itself be free of generating particles internally. Plus, of course, it had to have stuff like bar codes and automatic latching systems. It was an interesting problem, and it also involved the companion mechanisms to open and close the boxes and move the wafers into the process areas. This project was to stay on my "Tuh Do" list into 1984.

After we had pretty well developed the system concepts, I found out that another company, a spin-off from Hewett-Packard as I recall, had developed a remarkably similar system. There was no hanky-panky, just the not-uncommon case of different people recognizing the same problems and coming up with similar answers. Today the "SMIF" (Standard Mechanical InterFace) carrier system is in worldwide use. Ours isn't. Theirs was better.

And then another miracle occurred: On the Ides of March I was elected to be a Senior Member of the Technical Staff and bumped up from Job Grade 30 to 31. It was truly a surprise and I was pleased as punch. The grade of SMTS carried with it some swell perks in addition to the honor.

The first perk that I participated in required me to pay TI $500. Such a deal. But for the money I received a full-blown TI 99/4A Home Computer. The system had all the bells and whistles, including a modem so I could hook into the TI corporate computer from my house. The retail price of it was over $4000. They were not cheap, but don't forget, they couldn't do much either. But at the time they were the hottest things in home computers and I was ready to join the computer age. Soon I had the carload of boxes home, ripped open, and a real computer with printer, running in the study. There has been a computer in the same spot ever since.

The TI 99/4A was a lot of fun. Especially the speech synthesizer. Text typed on the screen could be converted to the spoken word with the push of a button. The words could also be modified with clever keyboard commands if you wanted to force a certain pronunciation, but all in all, it did a remarkable job. Its first public use in my house was at the end of our Bible Study class one evening. We all filed downstairs and crowded into the study and I pushed the "go" button on the pre-set computer. The 99/4A brought our evening to a proper close by reciting the Lord's Prayer. It got a nice round of applause, even if it did say "Who art in *heeven*..."

My dear friend Shirley Sloat related the tale of their 99/4A, which she and her husband Greg, also an SMTS, were using. Shirley admitted she spent an entire afternoon adjusting the synthesizer accents and inflections to say *"Shit!"* with four syllables, as pronounced in Texas. She's from Michigan and never could get it right.

Daughter Susan dropped by one evening and I was prepared to wow her with the synthetic speech by connecting through the modem to TI and downloading the day's edition of *TI News*, the TI on-line newspaper. Then I was going to have the computer read the newspaper out loud to her. Would she be impressed or what?

With the *TI News* on the screen, I hit the "speak" key and stepped back. The computer began speaking. "Asterisk asterisk asterisk asterisk..." I had neglected to notice that the *TI News* page was framed with a double row of guess-whats. We lost interest before the first row was through. She said she was really impressed by the demo anyway. She'd always wondered what those little stars were called. She was just trying to make me feel better.

The end of my TI 99/4A came when TI offered a buyback program. After using it for a couple of years I could return it and get my $500 back. Or, if I was clever, I could keep it and sell it to some sucker, er, computer enthusiast for whatever I could get and pocket the profit. Being a little suspicious of the market value at the time, I opted for the buyback. At the same time, TI was offering us SMTSs and higher technical ladder persons a free TIPC, the latest personal computer from TI. Now that was more like it.

We were given a budget figure of around $5000 and allowed to choose the options we wanted. It's hard to believe now how little computer you got for five grand, but it was a great deal at twice the price and it was strictly state-of-the-art at the time. With the options I chose I didn't get a hard drive, but I got plenty of other neat stuff. It really brought me into the computer age.

My TIPC ran for years, and I finally acquired a big hard drive for it—the 20 MB one—by a little horse-trading with Jim Sims. When I finally spent my own money on an IBM-compatible PC, the TIPC migrated to San Antonio where over the years it was passed down from daughter Bev and her husband Doug to their daughter Julia and finally to the end of the pecking order, my grandson Graham. Our families got a lot of use out of it and it's probably still out in their garage somewhere. Thanks, TI. That was a *great* idea.

Along with working on the clean carrier system and keeping my finger on the pulse of the vibration business, my group was still building Perkin-Elmer tops. As I said, the MMS III was finished and taken over by the Photomask Department in the latter months of 1981 and nightmares of The Project From Hell mercifully faded.

In our group's spare time we built a "Class Zero" room in the corner of the soon-to-be-ill-fated DMOS III cleanroom in the South Building. This super-clean area was to be used as our clean carrier test facility and we began equipping it for the particle business.

Somehow, in our additional spare time, we put together a TI-designed and built production robot, the TI-ER6000, and the TI A24D plasma etcher into an automated system. The robot loaded and unloaded wafers from the A24D. It didn't do it very well, but it was a start. We found out that robots don't work as well in real life as they do in science-fiction stories.

And then came another of those interesting interruptions. I was requested to go down with a small team to Corpus Christi and work with Richard Colley, TI Architect and world-class grouch. The TI Houston plant was preparing to build a fourth MOS (Metal Oxide Silicon) chip factory and wanted it to be laid out to comprehend future automation. This was a neat problem since we had no idea how we were going to automate anything. But our guess was as good as anybody's, and besides I liked to go to Corpus and work with Colley.

Colley and Associates had a low-lying, tree-shrouded building near the suburbs. It looked really cool like an architect's office should, and was augmented by an old Gulf service station in the rear, connected by a shady path. This ex-Gulf building, used for copying equipment and document storage, was clad in porcelain-coated steel panels and will probably last longer than the pyramids.

But the main building was a combination architect's office, aviary, and art museum. In addition to architecting, Colley collected rare parrots from all over the world; and his eclectic taste in art was displayed on every wall.

The parrots had their own room since they screeched pretty much all day. They were only faintly audible in our usual conference room. The walls in and around the conference room and most of the other walls in the building were covered with watercolors, paintings and prints from everywhere and of every style. As an amateur art lover I liked that a lot. In particular, several pieces by Frank Brangwyn, an English etcher, named caught my eye. He had a loose style of drawing and did scenes of everyday life that I found charming. It turned out that Brangwyn, who was also an architect, was a favorite of Colley's. Now, years later, I've got three Brangwyns of my own.

After a week of hard work and the usual verbal abuse from Mr. Colley ("Goddammit Ed, where the hell do you get those ideas?") I returned with a better idea of factory automation.

Several months later I incurred the wrath of Colley once again. The Houston bunch had asked me to do a floor vibration survey of the area proposed for the Perkin-Elmer printers before they did a factory layout. Good idea, since all floors weren't created equal, vibration-wise.

I loaded up my floor thumper and spectrum analyzer and headed to Houston. A day's work thumping and analyzing the results produced a puzzlement. There were areas very near the support columns, and directly on top of the "support wings" radiating from the columns, that were quite compliant. This meant that the areas would be prone to vibration if anything excited the floor, like a person walking across it. Not where you wanted to put a Perkin-Elmer.

I went into the pipe space to look at the floor from the bottom and couldn't understand the results I'd gotten. The support wings were physically big, stout solid concrete gussets between the columns and the bottom of the floor, four per column. It looked like a good design but it didn't work. And then I had an idea and borrowed a ladder from Facilities and climbed to the junction of the wing top and the floor above.

I fished a business card from my shirt pocket and stuck it between the two pieces of concrete. It slipped in without touching a thing. The floor and the support wings weren't connected. After pouring the floor, I suppose, it pulled away slightly as the concrete set and left a tiny gap.

If the floor was heavily loaded, I'm sure it would sag and rest on the support wings. But as it was, the floor was hanging out in the breeze and making like the proverbial drumhead.

I made the tactical error of calling Richard Colley and telling him about it. He considered my finding a direct and unwarranted attack on his skill as an architect and promptly dismissed the whole thing as a goddamn aberration of my twisted mind. As I recall, he said his goddamn concrete columns and goddamn support wings and goddamn floors would do no such goddamn thing, goddammit. Well, okay. I sent my report and recommendations to the head of the Houston layout team and didn't copy Colley. I thought the phone call was sufficient abuse.

Work progressed on the clean carrier system and some of the other bits of the automation puzzle. We installed a loaner Mosler "TeleLift" system in the Floyd Road building to try out a commercial system for moving material. Designed for banks, it was like a three-dimensional conveyor belt. A motor-driven box would scoot along a rail like a tiny berserk railroad engine. It was a nice system but would have been impossible to make it run clean enough for use in our production areas.

The other delivery device being considered was a neat mobile robot built by a group within TI. This self-propelled device, about the size of skinny desk, could drive itself around to a given location, energize the TI-ER6000 robot that was bolted on top, and retrieve or deliver cassettes of silicon wafers to a process machine or storage bin. We began working with Marty Wand, the mobile robot project engineer, about the middle of 1982. We did lots of factory layouts with the idea of using such a machine instead of people to move the wafer carriers between workstations.

229

And then, big surprise, Richard Colley called me. He said he was having trouble getting someone to calculate the resonant frequency of concrete columns. One of his ideas for reducing the vibration in the Perkin-Elmer printers and others of similar ilk was to put the printers on concrete pedestals that were poured on the vibration-free ground floor slab. The tops of the tall pedestals would be even with the manufacturing floor on the second level, but not be connected to it. As others before and after, Mister Colley had assumed I was a mechanical engineer. Although my bachelor's degree is in electrical engineering, I no longer took offense, and I was also equipped with plenty of mechanical engineering textbooks at my desk. I asked Mr. Colley for the dimensions and said I'd let him know.

This proposed support column, sticking up from the ground floor like a big concrete finger, would have a side-to-side resonant frequency like a tuning fork. If the natural resonance of this column was near the sensitive frequencies of the printer perched on top of it, it would be worse than useless. But how in the world do you figure out the speed a concrete tube would vibrate without building one and hitting it with a sledgehammer? And then I thought of Buford Baker.

Buford Baker, one of my favorite mechanical engineers of all time, was now deceased, but he had left me a wonderful legacy. Years ago he had given me a simple equation that was engineering magic. With it I was able to solve seemingly difficult vibration problems with great nonchalance, making only a scribble or two on my tablet during meetings. And what strength spring should be used to isolate a specific frequency? I could do that, stifling a yawn, while you waited. For the 0.03% of the readers who would actually like to know what the equation is, keep reading. Other readers may skip ahead several paragraphs without losing the thread, if any, of this story.

This equation, actually a close approximation, relates the sag of the spring in a spring-mass system to its resonant frequency. Everything else that looks important, like the mass and the spring rate, fall out during the derivation, leaving only the sag of the spring when the weight is attached.

The secret equation is (pause for effect) —

Sag (in inches) = **10 / f^2** where f is the resonant frequency in Hertz

The natural frequency of a concrete column can be approximated by turning the column sideways, like a uniformly-loaded cantilever beam, and calculating its natural sag, or deflection as they say in the ME books. Plug the deflection into Buford's equation and Voila! The answer! Just to be sure, I borrowed a tuning fork and measured it, and ran the same calculations on it. Instead of 440 Hz, I got 419 Hz, probably because the sag at the end of the beam is not quite the correct place. But as I "frequently" say, it was plenty close enough for this type of calculation.

Mr. Colley seemed pleased with the numbers that I gave him. The fact that he got any numbers at all was why he was pleased, and I knew I had just sewed up the concrete column frequency business. But then someone asked another question and Colley called back. How stiff is the column to the recoil from the moving parts of the printers? Putting my non-existent ME hat back on and digging out the equation for deflection of a point-loaded cantilever beam, I was able to give them the equivalent sideways spring rate at the top of the column. Colley was even civil to me. All because I had Buford's magic equation. Now you do, too.

The year motored onward with interesting progress on the clean carrier system, also called the "ultra clean" system. We had our "Class Zero" clean lab running in the South Building and were getting a smarter about particles and how to control them. We had worked with the manufacturing people and had done "what if" layouts of their production floors with mobile robots running around delivering wafers. And of course, I spent a fair amount of time messing around with the vibrations of floors and columns and Perkin-Elmer printers. As we ended the year, we had a few prototypes of our "ultra clean" carrier, which would hold two 125mm wafer cassettes. I thought they looked pretty good.

I thought they looked pretty good until early in 1983 when one of our new ultra clean carrier boxes was sent down for Richard Colley to see and critique. The carrier was returned with a one-word comment, "Pitiful!" That's pretty terse for a technical evaluation, but I soon received a page of specific negative comments from Mister Colley. I answered them one by one and never heard any feedback. We actually did know how to build a better clean carrier than Colley did. We'd been thinking about it and making tests for the better part of a year. Maybe it wasn't very pretty, but technically it was a good start. Pitiful, indeed! Just wait until you need another column calculation.

And to add to my personal frustration, I had lugged two pieces of my art, a.k.a. sculpture, over to Fort Worth in my VW bus and entered them in the TCU Juried Spring Art Show. My hernia-producing piece entitled Medieval Mechanism and the smaller Construction Number 22 both failed to make the cut and were summarily rejected. Both technical and aesthetic rejection in the same week? What's an artistic engineer to do?

But in the following year I entered my latest sculptural masterpiece, Mechanical Chair, in the D-Art juried show in Dallas and it was accepted. What a thrill! At the exhibit's opening night soiree I walked around holding a glass of wine with my pinky extended and sporting a nametag identifying me as "Artist." Wine never tasted so good.

This piece of "art" was an oversized chair constructed of heroic pieces of unplaned cedar pegged and tenoned together, replete with rusty chains and scary (I hoped) looking levers of unknown purpose. The title of Mechanical Chair referred to its supposed purpose as the predecessor of the electric chair. It was, like, for prisons that didn't yet have electricity. And as I watched during the evening, a young couple approached the chair and studied it after reading the description. She sat down in it and told her escort, "Okay. Work the lever!" He did, with vigor. She sat there a minute, then got up and said, "Humph! I didn't feel a thing!"

My career as an active "artist," using the term loosely, lasted quite a few years, and involved taking almost every evening art class that SMU and Richland Junior College had to offer. I had night classes at SMU doing woodcarvings and bronze castings that were both instructive and entertaining. One class in sculpture found me carving my locally famous (at least in my den) Adjustable Arm with Fish, which still hangs there. In the class a young woman attracted my attention by staying aloof from the rest of us *artistes* and single-mindedly dedicating her time to carving a huge piece of bois d'arc. This yellow stump was fully 18 inches in diameter and a similar height and weighed a lot. She spent her hours each evening with mallet and chisel, head down and unspeaking, making bois d'arc chips.

Towards the end of our four-week class, I decided to make a special effort to be nice to the lady, since no one had even spoken to her during the weeks we'd been working together. The rest of us had formed a gregarious group, verging on boisterous, and it seemed a shame that one member was being left out. So I walked over to her as she was carving, and said, "Hi, there! I'm Ed. What are you carving?" Just as I said, "What are you carving," I could see what she was carving. The bois d'arc cylinder had become a seated nude man, legs spread and curled around the outer edges of the wood, elbows atop his knees, intently studying his... well, take a wild guess what was in the center of the piece. "Uh, never mind..." I said and stupidly walked back to my adjustable arm and fish.

But about the time I had used up all of the SMU art classes, Richland put in a wonderful new art lab with lots of fine equipment. In addition to the usual potter's wheels, clay mixing machines and kilns, they had electric and gas welding stations and a complete set of woodworking power tools. That was for me. While the rest of the class messed with damp clay, I had the band saw running and was sawing up great pieces of wood for my artworks. Unfortunately, I was also making great clouds of sawdust, which had a tendency to settle on the clay pieces being lovingly shaped by the other class members. It made an interesting surface texture when the pieces were fired.

Mary Ruth and I had an opportunity to take a short vacation in April of 1983. She had seen an American Airlines offer of round-trip fares to a number of cities for $99. What a deal! She decided we should visit Boston, since it was the farthest city on the list and we'd get more for our money. I agreed. I also called Ken Martin, recently at TI but now at home in Woolwich, Maine. He had previously invited Mary Ruth and me to come up and spend a few days with him and his wife and see the Maine coast. This looked like a good time to combine two birds with one stone—the world's cheapest airfare and friends in Maine.

Ken was an employee of History Associates and had been under contract to do a book on the history of Texas Instruments. He and another History Associates person, Sally Simon, had worked in the TI archives for two years gathering information for the book. Sally, however, jumped ship, signed on with TI, and became the first manager of the TI Archives. Ken stayed with the book for another year, writing in his home in Maine.

I had gotten to know Ken as he interviewed TI employees while collecting information. We hit it off and our paths crossed a number of times. Enough times for him to invite us for a visit, and we took him up on it.

The book that Ken wrote about TI was finally finished, I think, in 1984. He had spent three or four years of his life on it but it was never published. I knew that it was not popular with Mark Shepherd. The sticking point, I had supposed, was a difference of opinion as to the cause of the TI down years. Although I saw a partial copy of it in the TI archives some years later, Ken said the only complete and edited copy was last seen in Mark's briefcase.

Later, after I had retired from TI and was working part time as Manager of the TI Artifacts Program, I heard that Mark was retiring and I wrote him a note about the book. I suggested that it would be a good thing to put it in the TI Corporate Archives for posterity, or maybe prosperity. He responded by sending up a box of computer diskettes to Ann Westerlin, Manager of the Archives. These diskettes reportedly held the only version of the final edit of the TI history book that Ken Martin wrote. And it was on eight-inch diskettes. Yes, at one time there were eight-inch computer diskettes.

These eight-inch diskettes and the data on them were compatible with only one computer system in the world—the TI "OOF" machine. The OOF, or more politely "Office Of the Future," was the first computerized secretarial station and it was way ahead of its time. No more typewriter for the lucky secretary who got an OOF.

But the problem at this time was that the last living OOF in TI had been retired from service the year before. The good news was that I had been keeping an eye on it and when they finally pulled the plug, I nabbed it and accessioned it into the TI Artifacts collection. In my part-time home in the North Building pipe space, I had the only OOF in the world.

We trucked the bulky consoles upstairs to Ann Westerlin's archive area and connected it up. Ann called an electrician to get the proper power for it, and since Ann had used one in an earlier life at TI, she soon had it was up and humming. The daisy-wheel printer was fitted with the last available ribbon and Ann began feeding the diskettes into the OOF. With the daisy-wheel printer putting out what seemed like four pages an hour, it took days to print out the "final" copy of the TI book. But it was eventually finished and the OOF returned to its place in my dusty technical mausoleum in the pipe space.

The one and only final copy of the history of TI is now under lock and key and accessible only by order of the head of the TI Legal Department. It was saved, but just barely. Now I'm not sure why, although a recent conversation with Mark Shepherd sheds some light to go with the smoke on the long and eventually disappointing tale of the TI history book. Mark said he spent a lot of time with Ken Martin on the book, and had really hoped it would be a fitting account of the exciting story of Texas Instruments. But he felt it was not. Simple as that.

To get back on the track, we had a fine visit in Maine with Ken and his wife and were personally toured up and down the coast. It was a good vacation and very reasonably priced, considering the distance.

In more or less the spring or summer of 1983, the name of our department was changed from Automation Systems Department, a.k.a. ASD, to Process Automation Center, or PAC. Since "PAC" could be pronounced as a single syllable, every time we referred to our department we saved two syllables. Work Simp lives.

Along with this change, although merely by coincidence, I received my very own office computer, a TIPC with color monitor. TI was fanning out computers for various levels of management and my time had finally come. The primary advantage of having a computer on your desk was to be able to send an "MSG," or TI internal company message, on the TI computer network without getting up and going across the room to the MSG terminal and printer.

This network that TI had built at great expense was worldwide. I could now, from the comfort of my office, send and receive messages from any TI plant in the world. This was not a trivial system in 1983. It was probably the largest computer network of any private company. We used it a lot.

If I sent off an MSG from my computer, I, and everyone else in the area, could hear the Cost Center's MSG terminal pin-printer crank up a few seconds later and print out my copy of the message. The whine of the printer was distinctive and penetrating, and I soon observed that a long whine was a full line of text, and a short whine was a heading or such. I set to work with a series of tests to verify my beliefs and soon found they were correct.

As a local test of subliminal hearing, I typed up a dummy MSG that had just the right lines of long and short text and sent it out to the printer. Sure enough, in a few seconds, the familiar sound of "CQ," the ham radio general call to anyone, came out of the printer in Morse code.

As I suspected, subliminal or not, Jim Sims and Bill Hudson jumped up in their cubicles and began looking for the source of the CQ. Both were hams, of course, and a ham can't turn off his CQ receiver even when he's working.

Another step forward in our group was the acquisition of a real cleanroom of our very own, following the demise of our borrowed space in the DMOS III area. The new one was an old and out-of-date "Horizontal Laminar Flow" room in the Semiconductor Building, but it was reasonably clean and we really needed a place to run particle tests. We moved our equipment in and set up for business.

This room had a ghostly problem that we were never able to solve. If we ran our particle counter to test the cleanliness of the air on a 24-hour run, invariably, in the darkest part of the night when it should be the cleanest, we would find a burst of particles on our counter print-out lasting for fifteen to thirty minutes. The best explanation we could come up with was that someone on the nightshift had a key to the area and had turned it into a secret love-nest. And instead of gowning up in the proper clean-room fashion, they were pulling their dusty clothes off. I could never find a volunteer to hide in the room all night to verify this.

Late 1983, a rare opportunity presented itself to me by way of my boss, Jim Moreland. It was recommended that I travel to the

TI semiconductor plant in Miho, Japan, and work with the engineers on the design of a wafer carrier box. They were in the throes of attempting to automate the handling of the wafers in their manufacturing areas, as were we. As was usually the case, there was not a common solution for a box that would work in both places, but an exchange of ideas would be useful for us both.

It was an interesting trip. Traveling in Japan by myself was difficult, and fortunately I was usually with either another knowledgeable US TIer or an English-speaking TIer from Miho. In Europe, most of the signs can be figured out satisfactorily, but not in Japan. I carried a book of paper matches with me at all times that had the name of my hotel and the city in kanji characters. When I rode the train back from visiting museums in Tokyo, for example, all I had to do was match up the name of the train station with the characters on my matchbook cover. Don't laugh, it worked. And it was just a short walk from the train station to the Dai Ichi Hotel where I was staying.

My most vivid memory of the Dai Ichi hotel was the size of the rooms. The words "broom closet" come to mind. The tiny bathroom apparently was molded out of single piece of plastic— sink, tub with shower and all. When I stood in the diminutive tub my hair was sucked up into the ventilator grill.

The group of engineers and designers I worked with spoke very little English, and I spoke no Japanese, other than "please" and "thank you," and of course, the most important phrase of all, "Ichi Kirin." But it was not too much of a handicap when designing something. We used up a lot of markers on the white board, and we got along fine. A question mark and an arrow drawn to a part were sufficient to make a pictorial explanation come forth. A "thumbs up" was the universal "Okay!"

They were a very hard-working bunch, and we regularly worked into the late evening before breaking for the day. And the next morning, new drawings of our day's discussions had been magically generated during the night.

This trip had come at an artistically awkward time for me. The TI Secretarial Council in Dallas had just put out the call announcing a logo contest. And me in Japan? Dirty pool. But I was not to be denied. I sketched out my entry and sat down at the ubiquitous MSG terminal near my work area in Miho and carefully typed up a funny-looking message for Mary Frost, Jim Moreland's secretary. I had briefed her by MSG earlier that I was going to send her my entry, and since she was a member of the Secretarial Council, she could submit it for me.

The MSG terminals had no graphics capabilities, but a person could make a crude picture by typing in asterisks all over the screen, and doing a kind of "Connect-the-Dot" thing. I finally finished it and sent it off to Mary. My proposed logo was a circle around a crossed quill pen and telephone handset, which I thought exemplified the versatility of the TI secretaries.

Later, back in Dallas, Mary showed me what she'd received. The circle wasn't a circle at all, but a squashed oval, with the quill pen and telephone similarly distorted. Seems that what you see on the sending terminal ain't what you get on the printout. But Mary understood and redrew my worse-than-crude representation and submitted it. Guess what? I won. Of course, you might have guessed that, since I would have left this incident out of the book if I'd lost.

The food in Miho varied from what I finally began calling UFOs to McDonald's Big Macs. Every few days I would have a "Mac Attack" and walk downtown to the McDonald's and have a Big Mac. They tasted just like home.

The UFOs, on the other hand, were "Unidentified Fried Objects." This reached a climax of a sort when the day finally came that I did not get an offer to eat out of the Miho plant at noon. It was with some misgivings that I followed Susumu Yamamoto, my Miho host and friend, into the Miho cafeteria. I could understand zilch about the menu, and finally loaded up with a UFO that looked pretty good, and a couple of other barely recognizable items.

As I left the serving area and headed for a table, I asked Susumu what it was I had chosen to eat. He looked around and apparently saw no information, and shouted my question over the serving counter. The reply brought a smile to his face as he translated cheerfully for me, "Fried pig heart!" The Japanese engineers of my acquaintance really did take a perverse pleasure in trying to con us Texans into eating strange things. Not wanting to give him any more pleasure, I had fried pig heart for lunch. It wasn't bad, actually, for fried pig heart.

1983 ended with the first delivery of clean carriers to the TI plant in Sherman, Texas, and the beginning of talk about automating DMOS 4. DMOS 4 was going to be the latest and greatest "front end" in TI, with a first attempt at the automation of material movement. We were going to be working a lot with them during the next year. Sounded good to me.

It was soon decided that a mobile robot would be used to fetch material to and from automated "warehouses" and delivery it to several ion implant machines. We would be involved in the automated warehouses, a.k.a. "autowarehouses" and the wafer carrier itself.

My group would design and build the clean carrier box for use in this partially automated manufacturing area, and Joe Ayers, head of the Process Automation Center's Sherman engineering group, would build the autowarehouses. They made a good choice in Joe and his talented bunch. Joe had paid his dues and then some in the automation business. I was of the opinion that Joe's group could build anything. Just give them the problem and some money and get out of their way.

All Joe had to do was to design from scratch a series of automatic warehouses of various sizes, build a prototype for delivery to Dallas in July, and five more by the end of the year. Near as I can tell, neither Joe nor any of his team batted an eye over the size of the job and the quick delivery. It was a really big job.

The autowarehouses, and there were six to be built, were monsters. Their purpose was to temporarily store the wafer carrier boxes between process steps on the manufacturing floor. The size varied from the smallest model that would be the input material buffer and designed to hold 20 of our carriers, up to the really biggie that would hold 188 carriers. Each of the two 188-carrier storage models was 38 inches wide by 10 feet 10 inches tall, and 17 feet 9 inches long. The total cost of these autowarehouses and spare parts came to about three-quarters of a million bucks. What a bargain. Other parts of the automation system, including three mobile robots and a small flock of fixed robots, added almost another mil. Automation is expensive. You can hire a lot of folks to carry boxes around for that amount of money. But it was a necessary step towards the future. Someday, someone was going to automate and it might as well be TI.

In operation, the mobile robot would drive alongside the autowarehouse and park itself accurately at the transfer port. A little computer-to-computer conversation would take place, and the warehouse would either take the carrier box from the robot and store it, or would deliver a requested carrier to be loaded aboard the mobile robot. And this had to be done without adding any particles to the air.

My group, or at least the ones not working on something else, began the design of a single 150mm cassette carrier box. We had worked with the engineers at DMOS 4 and developed the specifications. Mel Creps was our chief drawing-board person, and he completed the job by making the most complicated drawing I'd ever seen. This carrier box was to be made of molded plastic, and the drawing, as Mel knew and I didn't, needed to be interpreted by the tool and die makers who would make the mold. The mold was built by a local firm and cost $60,000. It was best to have really good drawings. The boxes, when they finally were molded and assembled, were perfect. Mel does nice work.

In March of 1984, I received a company-wide message from TIer Don Manus, beating the bushes for historical items for the TI artifact collection. I responded with two items I thought should go in TI's historical collection. The first was the HAL-9 computer, always dear to my heart, and the second was Jim Nygaard. I proposed having him stuffed and mounted and put on display. Don responded positively, and allowed as he would have Nygaard picked up for taxidermy before Jim retired on the 26th of May.

In the meantime, we had upgraded our particle test lab a lot. We moved from our spooky and antiquated horizontal laminar flow cleanroom to a new and spacious VLF, or Vertical Laminar Flow, space on the upper floor of the Semiconductor Building. It was much cleaner and we were getting well-equipped to run tests of all kinds.

My pride and joy was the receipt of a brand new PMS (no, it stands for Particle Measuring Systems) LPC-555 particle counter for our new particle-chasing lab. It could not only measure particles in the air that were as small as 0.12 microns, which is less than 5 millionths of an inch in size, but it could do it while inhaling the air at one cubic foot per minute. It was truly state of the art and cost $20,000. I had lobbied long and hard with Jim Moreland, and Jim with his bosses, to spring the money for it.

It was like waiting for Christmas until it finally arrived. I scrounged an oscilloscope cart to put the heavy counter on, and loaded them both in my car and headed for the Semiconductor Building and our new lab. Hot damn! This thing was an order of magnitude better than the LAS-X counters that we'd been using. I could hardly restrain myself.

I parked by the door, loaded the counter on the scope cart, and slowly and cautiously trundled my new baby to the building. As I crossed the threshold of the entrance, the front wheels of the scope cart snagged, and the already top-heavy load of particle counter pitched over forward, and my new $20K counter landed face-down on the concrete with a bang. Well, *shit!* I believe the instruction book said not to subject the counter to any shocks. I wondered if just one shock would be too many? One really hard shock on its front on the concrete, for example?

With the only other choice being to sit down and cry, I wrestled the counter back on the scope cart and continued the trek to our lab. A little work with the pliers fixed the bent sheet metal, and with a lot of nervousness and a short engineering-type prayer, I plugged it in and turned it on. I quickly ran it through its paces, and sonofabitch! It worked! Now *that* was one well-designed piece of equipment! I decided against mentioning this to anyone. Something about things that end well...

Remarkably enough, Joe Ayers and his crew designed and built the first autowarehouse, shipped it to our plant in Dallas, and had it up and running as a prototype by the end of September. It was a truly astonishing achievement. And it ran very well indeed, with a working demonstration of the mobile robot and the warehouse handing material back and forth as planned. And the other five warehouses were close behind in the final stages of assembly.

The only window of opportunity to move the three largest warehouses into the DMOS 4 building was rapidly approaching. And the window was on the second floor of the DMOS 4 building. The three biggest autowarehouses were so large that they could not be brought in through the shipping dock like any other self-respecting piece of equipment, but had to be hoisted up the outside of the building with a crane and stuffed in an upstairs window. The availability of the upstairs window would close before the end of the year as the construction proceeded absolutely pell-mell on the DMOS 4 facility. And of course, to nobody's surprise, Joe and his team met the deadline. That was an exciting day when the crane hoisted those babies in through the window.

The other piece of the autowarehouse puzzle, the software, was a joint effort between two of my guys, namely Dr. Jim Sims and Bill Hudson, and Billy Dollarhide and Joe Moore in Joe Ayer's Sherman group. Joe Moore was a new guy in their group, and worked literally night and day to get the software for the autowarehouses completed.

When I first met Joe in the summer of 1984, the whole Sherman group and I had gone to a cafeteria for lunch. We were eating around a big table, laughing and joking about the project, and someone said, "You can always tell software guys because they wear tennis shoes." We all had a big laugh, and then, one at a time, we each sneaked a peek under the table. Sure enough, Joe Moore was the only person wearing tennies. Then of course, we all fell on the floor laughing.

But the joke was on us. He was an excellent programmer and did yeoman's service on the autowarehouses, sneakers or not. I quote from a thank-you MSG that I sent Joe Moore in December, after the successful completion of the software:

> "...way above and beyond the call of duty...long hours...impossible deadline...work with the world's strangest software team of Sims, Dollarhide, and Hudson. You were truly running the hundred-yard dash with an anvil under each arm. You not only finished the race, but you did it in 8.9 seconds."

The automated warehouse project was an outstanding effort on the part of Joe Ayers and his entire wild and crazy group.

I had been signed up during 1984 to be the IDEA coordinator in the Floyd Road building. This was kind of fun, as it was an interesting "idea" by TI management to encourage far-out thinking. I could bestow upon a person, (almost) no strings attached, up to $25,000 for a far-out project. It didn't even need to be in TI's line of endeavor. It just needed to be radical. The only strings attached, kind of, were that the pursuit of this project shouldn't interfere with their usual daily duties. After word got out, I was treated with more respect, at least by those who thought they had good candidates for the IDEA money.

And I was also elected to serve on the Corporate Technical Ladder Nominating Committee for PAC. This group of engineers and scientists evaluated people who had been submitted by their bosses to become Senior Members of the Technical Staff or higher on the technical ladder. It was a tough committee to be on, since there were a lot of good and deserving candidates mixed in with a little not-too-subtle politicking.

To counteract all these wonderful appointments, I was dropped from the Semiconductor Patent Committee. But my service of seven years was plenty, and I'm sure a new face was welcome. Do you suppose Fred Bucy found out who first rejected his disclosure? Nah. He was too nice a guy to retaliate. And being cycled off the Patent Committee was a relief in a way. As much as I enjoyed it, it was time consuming and at times agonizing. My rewards for serving the time were the lasting friendships with the gang of patent attorneys at TI. Good work, guys.

Really on the down side late in the year was bad news from GEFE, the Government Equipment Front End, user of our first clean carrier boxes. They had found a serious problem. Wafers carried and stored in our boxes had a higher failure rate of "gate oxide integrity" than did wafers stored in conventional boxes. It was surmised that the failure was some subtle outgassing of the plastic sheet that was vacuum-formed to make the tops.

We later found out that the "pure" polycarbonate sheets that the carrier tops were made from had a few "standard additives," such as ultraviolet protection and smoke and flame inhibitors. The first clean carrier box was a chemical failure. We made sure the raw plastic that the new DMOS 4 carriers were molded from didn't have the additives.

To start off 1985 properly, Joe Ayers' group completed and delivered the three remaining autowarehouses to DMOS 4 on schedule in January. Meanwhile, we had been screwing molded plastic parts together like crazy and building the 700 flip-top DMOS 4 carrier boxes.

When I said "we" had been screwing boxes together, that was an Editorial We. Actually, Billy Flippin had been screwing them together. He called me one day and suggested that he'd be glad to pay half if I would just buy him an electric screwdriver. Good grief! I suddenly realized that he was screwing together 700 boxes with a hand screwdriver. Billy was almost embarrassed to bring it up, but he thought it would be a lot faster with a power tool. That, and he'd get fewer blisters. A quick trip to Sears solved the problem. I was only sorry that Billy had to bring it up. He never was one to complain.

In fact, in 1981, Billy Flippin had received my homemade "Misfit of the Year" award because of his strange conduct in the Feasibility Shop. In the words of the award,

He did not:

> *Use coarse and vile language, particularly to supervisors*
>
> *Go to sleep during finish cuts*
>
> *Take one-hour breaks 5 to 6 times a day*
>
> *Scratch unmentionable body parts in public*
>
> *Make uncouth and obscene gestures and noises to anyone within sight or hearing, especially supervisors.*

He was a true jewel of an employee. Quiet and productive, he really stood out in the shop. And we did deliver all 700 clean carrier boxes to DMOS 4, with Billy Flippin personally doing most of the work.

Sometime during this year I became involved with a problem in Lubbock concerning contaminants in the grease used in some equipment on the production line. How or why I got involved escapes me, but I began collecting data on the constituents of grease. There were certain chemical elements that would cause a product engineer to blanch and cross himself if mentioned in the same breath as semiconductor manufacturing.

I pestered grease manufacturers for parts per million data information on accidental or on-purpose trace elements and I sent grease samples to the TI chemical analysis lab to see what they contained. Soon I had a useful grease database, and before I knew it, I was known as "Mister Grease." I can't tell you how good that makes a person feel. I wanted to be an "expert" in something, although I had never even considered grease. But my lack of advanced degrees prevented them from referring to me as "Doctor Grease," which would have sounded a lot better.

1985 was the year that I delivered my own personal HAL-9 to Jim Lacy, manager of the TI Artifacts Program. We had a lot of fun with the HAL-9 at home, and the Christmas carols we had programmed into it several years before were still there. With core memory, nothing ever goes away accidentally. And Jim accepted it graciously and then sent it to the Smithsonian Institution along with other TI artifacts. Is that serious cool or what?

Towards the end of 1985, I was caught up in the work of a production improvement team and asked to look at automating the loading of wafers in an ion implant machine. An operator now loaded the machine by hand with a "vacuum wand" in a delicate and tiresome manner. After dealing a little with robots and in particular automatic loading of machines, I should have known better. But at the time, it seemed like a useful and worthy challenge. TI had a lot of ion implanters, and if each of them could be fitted with a robot loader, it would be a good business for us.

I looked around at available robots and ordered a Puma Model 260. It was a cute little booger, about the size and with the reach of second-grader. Onward, into automation!

One more year-end challenge came my way in addition to the ion implant loader, and this was so far off the wall it looked like fun.

245

The TI Facilities group, which was a large contingent of plumbers, electricians, carpenters and whatever, as a group, had just gotten into a heap of trouble in Sherman. I heard the story from several sources, although it was never officially verified, that Sherman Facilities personnel had been scheduled to do a simple weekend job. The installation of a new piece of equipment required that a moderate-sized hole be put through the concrete floor of a cleanroom.

On Sunday, several Facilities people walked into the deserted cleanroom and proceeded to jackhammer a hole in the concrete. The problem, as they found out the next day, was the cleanroom they had jackhammered in was full of Perkin-Elmer printers, where a speck of dirt can mean disaster. The printers were now coated inside and out with a thin film of concrete dust. The mirror systems in the optical paths were ruined. There was no way they could be cleaned and restored to health. With luck, the damage was only in the mid six-figure range.

The Sherman management was livid and set about to correct the problem. And the problem they wisely decided was not stupidity but ignorance. No one had ever told Facilities that you shouldn't jackhammer concrete openly in a cleanroom. So yet another task force was set up to teach the Facilities personnel about semiconductor processing in general and cleanrooms in particular.

I was elected, although I don't recall voting in the matter, to join the Facilities Training Task Force. The task force was of substantial size, with people from every department, all of whom were specialists or knowledgeable in some field touching on cleanrooms and semiconductor processing. After a number of meetings, a course plan was generated and the various bits of it apportioned out to us lucky soon-to-be teachers. We were totally responsible for setting up our lessons, including transparencies, handouts and demonstrations. This was not as popular with my boss as it was with Facilities, since we had to shoehorn this extra work into our already busy schedules. But it looked to me like a chance to do something useful as well as something I'd never done—teach.

My part of the Facilities Training Class was one and a half sessions, with a session being a full morning or afternoon. My first subject, a full session, was to present to the facilities folks what kind of processing it took to make computer chips. I covered briefly the equipment and how it worked.

The half-session I taught was about photolithography cleanrooms. These are the rooms where the Perkin-Elmer printers and similar machines reside in production. The emphasis here was on vibration and particle control. I brought my faithful Hewlett-Packard FFT spectrum analyzer and a vibration sensor and let class members jump on the floor and see how big a vibration signal they generated. I also had them practice saying hexamethyldisilazane.

From the beginning in 1986 through 1988, I taught my sections nine times. After a little stage fright in the first session or two, I really enjoyed teaching. Each class consisted of a couple of dozen Facilities personnel from all walks of life, and all well-trained in their specialties. They could have taught me a thing or two about their jobs, and it was fun to exchange knowledge. By and large, they were a lively group that put up with no bullshit and had minimal respect for authority, especially teachers. Maybe that was what made it fun.

The design and installation of the robot arm loader on the ion implanter was the big chore for me in first two-thirds of 1986. We sweated and worked and sweated some more. It was a difficult job, half again harder than I'd estimated. But it kept getting better and better and closer and closer to our goal. By August it was looking good.

It was looking good, that is, if you didn't count the ease with which the Puma 260, the cute little booger, rammed its metal fist through an entire boatload of twenty-five wafers, breaking every damn one. Now, why the hell did it do *that*? The really, really good news was that by sheer luck, the broken wafers were "pilot" wafers and not production wafers. Pilot wafers were wafers unusable for production and already written off as defective, and they were worth a lot less than the $200 to $1000 for each in-process production wafers.

Charles Gray, robot expert on the project, finally doped out the reason for this incredibly rude and what could have been incredibly expensive behavior. The robot had been halted in mid-stride for some reason, and then, when restarted, was told by its software to begin from a particular set of X, Y, and Z coordinates. It did as it was told, not knowing that a cassette of wafers was between it and the starting coordinates. It did what we do—it took the

shortest path, which happened to be right through the wafers.
Charles added a patch to his robot software.

But I gave Mohan Rao, Senior Vice President of Process
Automation Center, a good demonstration of the robotic loader on
the ion implanter, and contrary to Murphy, it ran splendidly. But the
end was near for our ion implanter loader. The cost of the loader was
too high, and the implanter we had designed the robot for was now
falling out of favor. The upshot was it wouldn't pay out quickly
enough for the fading usefulness of the implanter. Well, shoot.
Young lady, your job is safe. Keep loading those wafers with the
vacuum wand until we think of something else, and by then you'll be
a supervisor.

As the ion implanter robot fiasco, er, project, was winding
down, Jim Moreland and others had a really good idea. Dr. Ben
Sloan, our very own Vice President, had determined a year before
that the leading cause of equipment failure on the integrated circuit
production lines throughout TI was the loading and unloading of the
wafers. I could understand that. The process machines, complicated
as many of them were, worked better than the Rube Goldberg
mechanisms that put in and took out the wafers. So Jim Moreland
and Ben Sloan and Len Foster and probably some others decided to
see if they could fix the problem.

My favorite line in Shakespeare's *Henry VI,* "The first thing
we do, let's kill all the lawyers," paralleled their approach to the
problem. Except the first thing they did was to kill all the wafer
loaders. Big-time troublesome? Do what Work Simplification and
Shakespeare suggested—get rid of them, first thing.

So the new approach to a series of coupled process machines
was to strip them of their unreliable wafer handlers and place them
in a circle around a good robotic loader. It was called a Work Cell
and it was a good idea. Removing the loaders from six or eight
machines saved a lot of money and the money saved could be used
to buy a precision robot. It was an excellent trade-off.

My involvement in the Work Cell design was minimal,
although I found a block diagram of the control system that I had
done in October of 1986. But guys in my group were into it big time.
Jim Sims, Bill Reed, Bill Hudson and others worked full time on the

first prototype of the Work Cell. The Work Cell would eventually turn into an outstanding product.

I stayed on the fringes of the Work Cell project, doing some work on the spin motor and other odds and ends. In August I ordered a Puma 560 robot for the prototype. This big brother to the cute little Puma 260 had the reach of a Dallas Cowboy tight end and could slap a large knot on your head if you were unwary enough to get near it when it was energized. It was big enough to be dangerous and was never run without all the safety gear in place.

But 1986 did not pass without my Most Embarrassing Moment. Late in the year I had been working on the phone with a TI engineer in West Texas about static electricity problems and had finally arranged to fly out for a day and run a survey of his production area. I stuck the electrostatic voltmeter in my briefcase and hopped a Southwest flight to Lubbock. After renting a car I drove to the Lubbock plant, boogied in and asked the receptionist to ring up my friend. She sweetly replied, "He doesn't work here. He's in the Midland plant." Well, shit. Somehow I'd just assumed he was in the Lubbock plant and hadn't bothered to ask.

Soon, on the road again to Midland by rent car. It was only 115 miles, which was hardly an hour and a half drive in West Texas.

I was met in the Midland lobby by my friend, grinning from ear to ear. "Went to Lubbock? Haw! Haw! Haw!" He's damn lucky I ran his static survey for him. But the big mistake was confessing to Jim Moreland. He managed to bring it up every six months or so, just to keep me calibrated.

But Moreland was not one to rub things in. He was too nice a guy. Once I had sent out a formal message to a long list of engineers and scientists, inviting them to a conference "to *here* about the developments of..." Jim merely sent me back his copy with a penciled arrow pointing to "here" and his notation of "tee hee!" I thought that was humane, considering the faux pas.

But early 1987 brought the biggest embarrassment I ever heard of to an engineer at TI. Mary Frost Hagler, Jim Moreland's secretary, told me this story the day it happened. She was laughing so hard tears were rolling down her cheeks.

She had just gotten back from a meeting where a new engineer made his first big-time presentation to Tony Adams, Vice President of the Assembly and Packaging half of the Process Automation Center. The engineer was in the middle of his "foil" presentation when Tony asked him a question. It must have been a hard question, because the green presenter paused, stroked his chin in puzzlement, and said, "Hmmm!" As he said "Hmmm!" a booger blew out of his nose and landed on his transparency. Now, magnified a hundred times, it was projected on the screen for all to see.

If there was an end to this story by Mary, I didn't hear it. I was laughing so hard I nearly lost bladder control.

The process of photolithography, which is all the steps of putting a photographic pattern on the wafers, was getting more and more difficult as the patterns got smaller and smaller. The existing methods of coating the wafer with photosensitive material and developing the exposed pattern had no inherent feedback in the processes. To coat the surface, an accurate amount of photoresist was poured on the wafer, and the wafer spun at a certain speed for a certain time. Then, if everything were correct, the coating would be just the right thickness.

Developing the pattern on the sensitized and exposed wafer was the same deal. The wafer was spun at a fixed speed and hosed down with the proper developer solution at the proper temperature for a fixed length of time. Simple, certainly, but an open-loop process. Change any one variable and the result will change.

Several of us in Jim's group had been kicking these ideas around and we decided both processes could use feedback for control instead of dead reckoning. Feedback is good, and you may quote me on that.

The idea was that if we could somehow measure the thickness of the photoresist as it was being spun on, we could control the spin speed and make it just the right thickness to a lot less than a gnat's ass. We could no longer do things to a gnat's ass. A gnat's ass was way too big for our necessary precision.

And the same would be true of the developing process. If we could just find some way to tell when the wafer was properly developed, we could quit at just the right instant. Feedback would get around a lot of the problems of maintaining the exact conditions.

We decided to poke at these ideas a little, so we set up a small lab in an unused space on the bottom floor of the same old Floyd Road Building that was now the PAC Building. I scrounged a spinner and a couple of microscopes and other scientific-looking things and we were in business, kind of. Sam Wood, chemist, had just transferred into our group so he was to design and run the experiments. It got deep into physics, optics and chemistry before the year was out.

Our two goals for the year were to see if we could determine the film thickness of the photoresist "in situ" and the proper completion of the developing process by means of an "end point detector." This was good since I always wanted to do something 'in situ." It sounds so cool to stick a Latin phrase in your reports. "...*in situ*..." Cool!

My year was interspersed with airflow measurements and corrections on the Work Cells. In particular, one of the new Work Cells in the Houston Bipolar 4 facility didn't want to straighten up and flow right.

I headed to Houston with my one-and-only "Mark IV Airflow Indicator." It was a collapsible pocket pointer with a feather on the end of a foot-long fine thread. Don't laugh. It was ten times as sensitive as the best hot-wire anemometer and cost about $2.00. Just a tip—Plymouth Rock feathers work best.

I was gowned, booteed, gloved and hooded up to my eyes, the usual mode of dress in cleanrooms, and dangling the feather on my secret weapon around the recalcitrant Work Cell. I had about figured out the problem when I noticed one of the line ladies eyeing me suspiciously. When she strode purposefully towards me I retreated a step, as I was on her turf and she outweighed me by half. She stopped a foot from my face, put her hands on her hips, and told me in no uncertain terms, "You ain't gonna catch *nothin'* with that thing. It ain't got no *bait* on it!" I cracked up on the spot, which probably caused me to emit a burst of particles.

The year 1987 ended on a pleasant note. I was over in the Central Research Building working with a couple of my favorite scientists, Bob Bowling and Wayne Fisher. We had a three-way discussion going about the best way to clean particles from the surface of wafers. The big particles could be removed with various types of scrubbing, but the laws of physics ganged up on you when the particles were less than a micron. Since a micron is about 40 millionths of an inch, we're talking here about some tiny dirt.

We began kicking around ideas about somehow making the tiny particles bigger so we could sweep them off. One thing led to another with each of us chipping in a new piece of the solution. In the matter of an hour, we had roughed out U.S. Patent 4,777,804, to wit, "Method And Apparatus For Easing Surface Particle Removal By Size Increase."

This was the easiest invention I ever worked on. I think Bob wrote up the disclosure but I got just as much money as he did. It was a neat idea and almost perfect except for one detail. On the back of the patent's Sheet One is Figure 4, a drawing of the proposed machine. For Pete's sake, did they have to draw the spray nozzle to look like a penis?

TI had begun making regular retirement overtures to us older employees. Every time one came out I looked it over optimistically, but none seemed right. However, the idea had been planted that if I got tired of working, I could retire. It was a revolutionary idea to a person who had been working since a teenager. Whether Mary Ruth and I would starve to death was the second question. But with all our kids out of college and gainfully employed, retirement had actually become a viable option. But not quite yet, I decided. I began 1988, my 37th year at TI, at the age of 58 which was just a bit too young. And besides, I liked working at TI. I was making a good salary and had projects to work on. Onward through the fog, I decided. But having a retirement offer waved in front of my nose was very distracting. The possibility of retiring was getting closer. Much closer.

About half of my group was working full time on the Work Cell project. The idea of stripping the wafer-handling mechanisms from the process machines and grouping them around a robot arm looked like a winner. It was more reliable and the up-time, or the time of actually doing production work, was higher than ever in the first experimental Work Cells in the TI Sherman plant.

The TI Sherman plant was different from the other TI semiconductor manufacturing facilities. The Sherman bunch was independent of the rest of TI. They made their own decisions about which equipment to buy and how to run their lines, and they did a good job of it. And the best thing was that they believed the Work Cell was a good piece of machinery. They were the first to jump on the bandwagon, even before there was a bandwagon, and support our group in PAC on its development.

The success of a new manufacturing tool or system depended strongly on the reports from its first users. If your machine was cranky or hard to maintain, word would get out and you might never sell another one. But, in the case of Sherman and the Work Cell, they liked it and supported us and gave the rest of TI positive reports on it. Sales began picking up and it was on the road to becoming a good product for PAC.

My only personal work with the PAC Work Cell was in the internal air flows. That was because I owned the Mark IV Air Flow Indicator. The contamination of the wafers from particles in the air was highly dependent on the way the clean air flowed through the system, and in the air pressure inside the walls of the Work Cell. My feather on a string and I traveled around adjusting air flows in Work Cells all year.

My job as IDEA coordinator took up a little of my time, and I "funded" four far-out projects during the year, much to the delight of the recipients. But none of them were the lottery winners that TI was secretly hoping for.

I taught three more of the Facilities Training classes which by now were pure fun and no hassle. They even gave me a swell plaque to hang on the wall with my likeness burned in walnut by a laser. And for another fun sideline, I was a session moderator for a rare US visit from the famous Professor Ohmi from Japan, discussing the design and fabrication of high-yield semiconductor manufacturing facilities.

But mostly during 1988 I struggled from one thing to another, without a good solid project I could get my teeth or my slide rule into. It was kind of like eating leftovers, and just not very satisfying. Could it be that the merry-go-round was finally running down?

I was "invited" by Jim Moreland to join a team to write a proposal for government funding. It was for the MMST project. This was a drag because it had to be written in a precise and structured manner and in the proper tense and with correctly spelled words. How can anybody be creative with such stifling rules? I duly wrote my section, and rewrote it a time or two, and it was a real "spinach job." I had to eat it in order to sustain life whether I liked it or not. I've mercifully forgotten what "MMST" stood for.

The work with Sam Wood on the developer end-point detection apparatus was interesting and frustrating at the same time. Our homemade lab was ill-equipped and the problems complex and difficult. Progress was slow when at all. It was like swimming in molasses: an awful lot of work with little progress. Our work never approached anything resembling a viable production tool. The most useful product might have been the three patents that Sam and I jointly produced in this field. Let's just call them defensive patents.

Len Foster and I turned in two other patent disclosures in 1988. Len ran the mechanical group under Jim Moreland and I ran the mostly electrical gang, the "Process and Applications Engineering Section." Len and I collaborated on the design of vacuum cassettes for storing and moving wafers in and out of vacuum chambers and managed to get two patents out the work.

My best decision in 1988 was to get the old Super CAT gang and spouses together for a party and reunion. My invitation, titled "Gathering of the CAT People" went out to the guys who designed the Super CAT transistor test system in 1961. I mentioned in the heading of the invitation that we should be "...past the punishment phase. Besides, Shepherd can't hurt us now." It had been 27 years since that wonderful spring of 1961 when it all suddenly came together into a truly significant machine. Once again, the team:

Jim Nygaard – Electrical engineer, Aggie, and genuinely fearless leader
Jim Ricks – Electrical tech, both smart and smart-ass
Bobby Howell – Electrical tech and utility infielder
Bob Chanslor – Mechanical tech, machinist, watchmaker, fiddle player, and world-class humorist
J.C. Baggett – Mechanical tech, machinist and creative trouble-maker
Troy Moore – Mechanical tech, machinist and youngest by far

Lee Blanton – Electrical Engineer and generator of good and
    useful ideas
Harry Waugh – Electrical Engineer and knower of everything
Jim Anderson – Mechanical draftsman and frequent victim
    and me, Ed Millis – Electrical engineer and now book writer

And The Gathering of the CAT People was at our house on
Saturday evening, the 12th of November 1988, except for Bob and
Frances Chanslor. Bob picked a lousy day to feel bad, and we missed
him a lot. He was one of the long-time key players in Jim Nygaard's
Floating Menagerie of Technical People. We had a great time, a lot
of which was spent picking on Nygaard. He couldn't hurt us
anymore either.

The year mercifully slipped away and 1989 began about like
1988. Nothing I could get excited about and nothing that I could see
in the future that looked promising. The merry-go-round that I had
jumped on so bright-eyed and eager in 1950 was finally coasting to a
stop. It was time to go.

The chance came sooner than I expected. On January 27,
1989, there was a TI News Bulletin titled "Voluntary Reduction
Program Begins Jan. 30."

My age, which would be 60 on March 3, and my years of
service easily met the requirements for early retirement and the deal
looked pretty good. The main thing it included was a continuation of
the TI medical benefits at a reasonable cost. If I had any lingering
reservations about retiring, they suddenly vanished. It was time and
I was ready.

Mary Ruth agreed that it looked like a good retirement
package, but she couldn't say the word "retirement" without
gagging. In the fall of 1988 I had written up a "rules of engagement"
for Mary Ruth and I to discuss, about what we would do and not do
if and when I retired. She had later gone through the document with
her red pencil and changed the word "retirement" to "new life"
everywhere it appeared. She couldn't stand the idea of being married
to a "retiree." I didn't much care for the title either, and suggested
that I could be a "consultant" instead of "retiree." That sounded a *lot*
better, and Mary Ruth agreed that I could be a consultant as I sat
around the house and got in her way in the kitchen. Yeah, you
betcha.

The "Acceptance Form" was passed out to interested parties on January 30, and mine was filled out and returned on January 31. I wanted to be sure to get it in before the March 15 deadline.

And so, on the 31st of March 1989, I retired from Texas Instruments after an official 37 years of service. At the going-away party in the PAC Building, I found myself standing in a long line of retirees waiting to give the audience a few parting words of wisdom. I think we were allotted 30 seconds each. I said, "Thanks! It's been a great thirty-four years at TI, and thirty-four out of thirty-seven ain't bad!"

And it had been great ride on the TI merry-go-round. How could I have chosen a better career path? To fly in a PBY patrol bomber? To send my designs to the bottom of the ocean? To build the CAT and Super CAT? To play Bach in a New Mexico cave on company time? But most incredible of all, to be in the midst of the birth and blustering childhood of the semiconductor industry from the point-contact transistor to the 64-meg DRAM? My cup truly runneth over.

the end

Index

L

M